PCM PUBLISHING
LaGRANGE, OH - MARSHALL, NC

PUBLISHERS OF:
PETROLEUM COLLECTIBLES MONTHLY
AND OTHER INFORMATION FOR THE PETROLEUM COLLECTIBLES HOBBY
PRESENTS...

AF737539

GUIDE TO GASOLINE LOGOS

WAYNE HENDERSON
AND
SCOTT BENJAMIN

First published in 1997 by PCM Publishing, 411 Forest Street, LaGrange, Ohio, USA, 44050

Copyright 1997 by Wayne Henderson and Scott Benjamin.

All rights reserved. With the exception of quoting brief passages for the purpose of review no part of this publication may be reproduced without prior written permission from the publisher.

PCM Publishing is a partnership governed under the laws of Ohio and the United States of America. PCM, PCM Publishing, Petroleum Collectibles Monthly, and the logotypes and masthead design used are certified trademarks of the partnership and may not be used without permission.

The information in this book is true and complete to the best of our knowledge. All information and recommendations are made without guarantee on the part of the author or publisher, who also disclaim any liability incurred in connection with the use of data or specific details.

We recognize that some words, designations, and the thousands of trademarks shown are property of the trademark holder. We use them for information and historical research purposes only. This is not an official publication.

PCM Publishing's books, along with others created by the authors, are available from PCM Publishing. Some bulk discounts may apply to qualified parties. Contact PCM Publishing at the address shown above for more information or call (216) 355-6608 or fax (216) 355-4955. Credit cards accepted on phone and fax orders.

NOTE: NEW AREA CODE (440) AFTER SEPTEMBER 1997

Created at Marshall, NC, spring 1997. Printed and bound at Munson Printing, Red Wing, MN, May 1997.

Dedicated to my children, Jason Carl Henderson and Christopher Wayne Henderson, and to my parents, Carl and Jackie Henderson, without whose continued loving support this book would not have been possible.
Wayne Henderson...April, 1997

To learn more about this hobby see page 137 about Petroleum Collectibles Monthly, a monthly magazine devoted to petroleum collectibles.

The authors wish to thank the many people who, over the past 25 years, have contributed information and logos to help make this book possible. Contributors include Chris McKee, Bob Scism, Dave Leach, Walt Wimer, Jr., George Kraus, R. V. Witherspoon, Sam McIntyre, Thornton Beroth, Larry Stancil, Chip Flohe, Patrick and Barbara Shimmin, Jim Masson, Noel Levy, Bill Blair, Harley Allen, all of the many contributors to Petroleum Collectibles Monthly, and those who have traded and sold to me the many logo items, maps, matchbook covers, and other items over the years.

Special thanks again to those who, without their help, this book would not have been possible...Chris McKee, Bob Scism, and Walt Wimer, Jr.

Special thanks, too, to Happy and Chip Flohe, for their effort in making the 1993 edition of this work a success, paving the way for this more complete work.

Special thanks goes to my mother, Jackie Henderson, and my oldest son, Jason Henderson, for their effort in all the scanning, typing, and creating the book layout.

Thanks also goes to the hobby, who's patience over the past years has seen this project through. The friendship of each and every one of you makes it all worthwhile.

INTRODUCTION;

Congratulations on your purchase of Guide To Gasoline Logos, the most complete gasoline brand marketing history ever created. Inside you will find a series of charts that detail the eras in which various gasoline and oil marketing logos appeared. Combined with the text given on each corporation and company, the charts will give you a picture of the various brandnames and images used by petroleum marketers throughout this century.

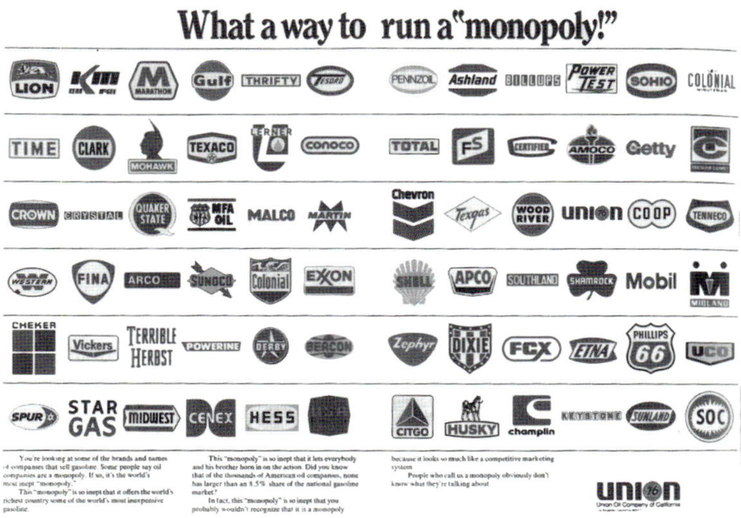

This guide has been 22 years in the making. In 1975 Union Oil, in an effort to defend itself against public outcry against major oil companies, published the ad shown here in major magazines. The ad, entitled "What a Way to Run a Monopoly" was created to point out that the oil industry was not a conspiracy and that indeed hundreds of independent marketers existed across the nation. Of the 72 brandnames pictured in the ad, only 41 remain in use under the same basic ownership today. The constant buyouts of smaller marketers by large conglomerates has, since the ad was created, greatly reduced the choices available to consumers today. Yet many independents remain strong, and jobber marketers, either branded or unbranded, remain the backbone of petroleum marketing in this country.

The "Monopoly" had a different affect on me than that which was intended. At age 13, and already a collector of petroleum memorabilia, I decided that I would someday create a chart that would show not only the emblems of current marketers, but indeed would show their images throughout their history.

The first project was two color charts printed in a high school graphic arts class. Although primitive, the charts gave the few interested parties around in 1978 a look at gasoline logos over the years. One chart was created for the Standard descendant companies, the other for the remaining majors. They were horribly inaccurate, but interesting nonetheless. The project remained on hold for many years, as the petroleum collectibles hobby matured, and I continued to gather logos and information. In 1993 we assembled by hand the first creation of "Guide To Gasoline Logos". Despite being hastily put together and a photocopy effort, the book was an incredible success. Three complete printings were sold out in less than six months. Since then, many of you have requested another printing, but the quality of the original did not justify a reprint. Better to simply recreate the book. One of our publishers, Motorbooks International, purchased the rights to the book in 1992, yet canceled the project before it was completed. Since that time the effort has been made to again present this work to you in some format. With the success of our monthly magazine, Petroleum Collectibles Monthly, it was decided that we would form PCM Publishing and release several books on our on. You are holding a copy of our first effort, one in which we are rightly proud. We hope you enjoy this book, as well as all our others, for many years to come.

COLLAGE COVER OF THE 1993 BOOK

Be sure to read the section entitled "How To Use This Book" as without reading this first, some features of the book will be difficult to navigate.

TABLE OF CONTENTS

HOW TO USE THIS BOOK

Guide to Gasoline Logos begins by showing the logos of each corporate assemblage of 44 major oil companies; followed by one section of industry symbols and four sections of independents, grouped geographically. Each company owned in whole by that corporation at any time in its history is grouped under the corporate heading. Several brandnames appear in two or more groups, as corporate ownership changed through the years. A complete alphabetical index follows the independent section. The omission of any independent from that section is not an indication of any disrespect intended for that operation, but rather results from a lack of logos and information. If you are in any way connected with any company not listed, feel free to contact us for inclusion in any re-creation of this work in the future.

The book is designed such that anyone having an item of interest displaying an oil company logo can reference the brandname in the index in the back of the book, and then refer to the appropriate section for a look at a representation of logos used by that company. The dates appearing directly under each logo tell the APPROXIMATE time each logo was used in official capacity. Your item then can be dated, within a range of years. Reading the text accompanying each section can give further information about that company and its collectibles.

The limitation of this book is the corporate policies on image standards. No petroleum marketer had exacting image standards in its earliest years, and their marketing images varied, albeit slightly, with each reproduction. What does this mean? Simply that the graphic image supplied to a signmaker to create an identification sign, may have been a professionally drafted image. But the graphic image then furnished to the makers of a pump sign, or the company that lithographed an oil can, or the company that furnished a road map may have been hurriedly hand drawn, with great creative license given to the end designer. As a result, you will often find only a variation of the logo appearing on your item appearing exactly in this book. Each image, though, will help you establish the approximate age, which should be acceptable to most collectors. As companies developed image standards, image manuals were issued and logos became much more uniform. Later items are much easier to date. The relative age of institution of these standards has more to do with the age of the company, though, as image standards are a product of company maturity, not the calendar year time periods of the past century.

Another limitation is the fact that most trademarks, names, and images are phased into and out of use. Exact dates of introduction are impossible to pinpoint, as many introductions took place over a period of several years.

Finally, read the text. It will greatly assist you in locating more information about the companies that interest you.

Amoco

STANDARD OIL COMPANY OF INDIANA

1912-1939 | 1926-1946 | 1934-1945 | 1945-1946

STANDARD OIL COMPANY OF NEBRASKA

1910s-1925 | 1925-1946

AMERICAN OIL COMPANY, BALTIMORE, MD

1915-1923 | 1923-1932 | 1932-1954 | 1954-1961

PAN-AM SOUTHERN CORPORATION NEW ORLEANS, LA

1923-1919 (GREEN/CREAM) | 1929-1938 (BLUE/WHITE) | 1938-1945 | 1945-1946 | 1946-1956

MIDWEST OIL CO. CASPER, WY

PRE-1929

ARRO OIL CO. LEWISTON, MONTANA

PRE-1929

UTAH OIL REFINING CO. SALT LAKE CITY, UTAH

PRE-1929 | 1929-1930

Amoco

1946-1961 | 1961-1971 | 1971-PRESENT

LOGOS USED WITH CONSOLIDATION OF STANDARD AND AMERICAN BRANDS

1961-1971 | 1971-PRESENT

AMERICAN BRANDNAME (REPLACED AMOCO AND UTOCO, 1961)

1956-1961 | 1961-TRANSITIONAL | 1961-1971 | 1971-1974

1930-1945 | 1930-1945 | 1945-1946 | 1946-1961

Amoco

TRUE'S OIL CO., SPOKANE, WASHINGTON

1920s-1930s

1970s SECONDARY BRAND

OMEGA OIL CO.
DAYTON, OHIO

1970s-1980s

1982-1996

SING OIL COMPANY
THOMASVILLE, GEORGIA

1935-c1960

1960-1980s

BRAND AREAS, 1961

AMERICAN PETROLEUM CO.
DAVENPORT, IOWA

1940s-1959

EXPORT

1946-1961
EXPORT

1971-PRESENT

AMERICAN OIL CO, DALLAS, TX

1930s

****AMOCO CORPORATION:**
****STANDARD OIL COMPANY OF INDIANA - CHICAGO, IL**

The American Oil Company was founded in Baltimore in 1910, refining and marketing kerosene. In 1915 they introduced a gasoline/benzol blend they called "AMOCO-GAS". Without crude oil production of their own and forced to buy on the open market, they operated with the possibility of shortage at any time. This was remedied in 1923 when a supply agreement was signed with Pan American Petroleum, based in New Orleans, giving 50% interest in American Oil to Pan-Am in return for the assurance of crude oil supply. Pan American, primarily a production company with trace marketing in several southern states, began its operations in 1916 and by the time of the 1923 merger was already an international company. In 1925 Standard Oil of Indiana purchased a controlling interest in Pan-Am, in the wake of Pan-Am's involvement in the "Teapot Dome" Scandal. In 1933 Pan-Am, now a subsidiary of Standard of Indiana, purchased the remaining 50% of American Oil and converted the Pan-Am marketing facilities in Georgia and Florida to the Amoco brand, retaining the Pan-Am name in Alabama, Louisiana, Mississippi and Tennessee. Arkansas and Kentucky were added at a later date, while the Amoco brandname was used in all states from Maine to Florida as well as West Virginia and Ohio. Standard bought the remaining interest in Pan-Am in 1954 and in 1956 the Pan-Am name was replaced by Amoco in the six southern states. At about the same time the Amoco brand had been introduced in Texas. While the Eastern seaboard states and Texas retained the old red and black Amoco oval, marketing in the former Pan-Am states was under an Amoco torch and oval logo similar to the ones used by Standard and affiliates since 1946. In 1961 the entire Amoco marketing area was converted to the "American" brandname and pumps were re-imaged. In 1971 the Amoco brandname was reintroduced for motor oils, and in 1974 was adopted nationwide for gasoline. Although the Standard brand still holds on in some areas of the Midwest, the Amoco name appears in over 30 states today.

****STANDARD OIL CO. OF INDIANA - CHICAGO, IL:** Formed in 1889 as a refining and marketing division of the Standard Oil Trust, Standard of Indiana had established an extensive marketing organization by the time of the 1911 breakup of Standard. Primarily a marketer, Standard established an extensive network of gasoline stations throughout the Midwest. Standard purchased Pan-Am in order to gain crude oil production, as is explained in detail in the Amoco section above. Midwest Oil Refining, listed below, was another producer purchased by Standard in order to keep its extensive refining and marketing operation going. In 1961 Standard shifted all domestic marketing to the subsidiary American Oil Company. Stations in former Amoco, Pan-Am and Utoco territory were rebranded American, while stations in the Standard territory remained Standard. In 1974 the Amoco brand replaced American and began to replace Standard, but many stations retain the Standard brand today. Standard Oil of Indiana became Amoco Corporation in 1984.

****STANDARD OIL OF NEBRASKA**: A marketing operation founded in 1906 and long supplied by Standard of Indiana, they merged with Standard of Indiana in August, 1939. Stations retained the Nebraska identification until after WWII.

****MIDWEST REFINING CO. - CHEYENE, WY:** Midwest was a Rocky Mountain producer/refiner when Standard of Indiana purchased controlling interest in the company in 1917. In 1928, Midwest entered the gasoline market, purchasing Arro Oil and Refining of Lewiston, Montana, Aero Oil Co. of Cheyene, Wyoming and 75 gas stations in Colorado and New Mexico from Vickers. The Midwest brandname replaced the individual brands. In April 1930, Standard of Indiana purchased the entire Midwest marketing and rebranded the stations as Standard. The Midwest subsidiary, Utah Oil Refining Co., having ventured into gasoline marketing, continued to market under the Utah Oil brands.

****ARRO OIL CO. - LEWISTON, MT:** Purchased by Standard's Midwest subsidiary and rebranded Midwest in 1929. See more details above.

****UTAH OIL REFINING CO. - SALT LAKE CITY, UT**: Founded as a subsidiary of Midwest Refining in 1928, the company marketed motor oils under the Vico name and gasoline under the Pep 88 brand in Utah and Idaho until after WWII. The "UTOCO" brand was introduced at the end of WWII and gradually replaced Vico/Pep 88. In 1961, when all of Standard's U.S. marketing was assigned to the American Oil Company subsidiary, the American brandname replaced Utoco. In 1974 Amoco replaced American.

****TRUES OIL CO. - SPOKANE, WA:** An independent marketer with stations in Washington, Trues was purchased by Standard in 1959. Stations remained under the Rainbow brand until after 1961 when most were rebranded American. Standard continues to market under the "Rainbo" name as a secondary brand.

****OMEGA OIL COMPANY - DAYTON, OH:** Amoco subsidiary brand operating in Ohio. Rebranded to Amoco in 1996.

****SING OIL COMPANY - THOMASVILLE, GA:** Independent marketer purchased by Amoco in the early 1990s.

****AMERICAN OIL COMPANY - DALLAS, TX:** Independent purchased in 1960 when the American brandname was established nationwide.

****AMERICAN PETROLEUM - DAVENPORT, IA:** Independent purchased in 1960 when the American brandname was established nationwide.

ASAMERA

**ORIENTAL REFINING,
DENVER, CO**

1930s-1950s

1960s-1970s

**ANDY-DANDY OIL COMPANY,
FT. COLLINS, CO**

1955-1964

**PAT GRIFFIN OIL COMPANY, FT. COLLINS, CO
GALLAGHER OIL COMPANY, BOULDER, CO**

1958

1958-1964

1964-1982

1982-PRESENT

GASAMAT SECONDARY BRAND

1962-1964

1964-1970s

****ORIENTAL REFINING - DENVER, CO:** Oriental was a long time Denver based refiner-marketer, dating back at least into the 1930s. They marketed under the Oriental brandname in several Rocky Mountain states. In the 1970s they became part of the Canadian conglomerate Asamera, which also purchased the Pat Griffin Oil "Gasamat" operations. In 1982 Asamera sold their U.S. petroleum assets, and the Gasamat brand is today operated by former Asamera executives.

****ANDY-DANDY OIL COMPANY - FT. COLLINS, CO:** In 1955 two Ft. Collins businessmen invented a coin operated gas pump and formed the Andy-Dandy Oil Company to set up coin operated gas pumps in and around Ft. Collins. They sold out their operations to Pat Griffin Oil Company in 1964.

****PAT GRIFFIN OIL COMPANY - FT COLLINS, CO:** Pat Griffin, longtime Phillips 66 distributor, saw the success of an upstart local competitor, Andy-Dandy Oil in the operation of self service gas stations. He converted several stations to use coin-operated pumps and branded them Gasateria. Conflicting with the existing Indianapolis based Gaseteria brand, he quickly changed the name to Gasamat. By the early 1960s the coin-ops had given way to self-service stations in which the cashiers actually lived. Payment was made at a drive up window with a forced traffic pattern and the attendant never had to leave the comfort of his living room! The resident-attended Gasamat (and later Gas-Mart, a secondary brand) survived into the 1980s, after Asamera had purchased the operation. When Asamera sold their U. S. petroleum operations, the Gasamat operation was sold to a former Asamera executive, who continues to operate about 30 stations today from a headquarters in Boulder, CO.

Ashland

****ASHLAND OIL and REFINING-ASHLAND, KY**

Ashland Oil was founded in 1924 as a refining division of the Swiss Oil Company of Lexington, KY, which had numerous ties to National Refining. Ashland began marketing gasoline under the brandname "Pepper", with the gasolines dyed red and green as well as clear to market the famous Red Pepper, Green Pepper and White Pepper products. Stations under the Pepper brandname were operated in an area of about 250 miles around the Ashland, KY headquarters. Branded gasoline marketing was all but discontinued during WWII, and after the war when retail gasoline sales resumed, the brandname Ashland was chosen, Ashland Oil having replaced the corporate name Swiss Oil in about 1935. Ashland has grown through the many mergers listed below, with branded Ashland outlets found within about a 200 mile radius of Ashland, KY and with company owned and operated stations, mostly under the SuperAmerica brand, throughout the central and eastern United States.

****TRI-STATE - KENOVA, WV**: Tri-State, an independent refiner serving much the same territory as Ashland did in its earliest day, merged with Ashland in 1930. The Tri-State gasoline brands were retained until the early 1960s as an Ashland secondary brand.

****INDEPENDENT QUALITY SERVICE:** Independent Quality Service was an Ashland secondary brand operated for independent, unbranded dealers from the mid 1940s through the 1960s.

****NATIONAL REFINING - CLEVELAND, OH**: The National Refining Company was a large refiner/marketer based in Cleveland, Ohio and founded in 1882. Marketing was concentrated throughout the eastern and central states with a subsidiary in Canada that was spun off as a separate company in the 1930s and survived until its 1964 merger with Shell. National marketed the famous White Rose Gasoline and for many years their boy and slate logo was one of the best recognized trademarks in the country. After WWII National was greatly weakened in the marketplace and the major portion of the company merged with Ashland in 1950. National White Rose stations were rebranded Ashland in about 1957. Unique in the gasoline business, the White Rose brandname was franchised to Mid-West Refineries of Muskegon, Michigan in about 1946 and although Mid-West merged with Leonard in 1955, the White Rose name remained in use in Michigan through the early 1970s.

****AETNA OIL - LOUISVILLE, KY:** Aetna Oil, dating from the 1910s, marketed through service stations in central and western Kentucky and southern Indiana. Aetna merged with Ashland in 1950. The Aetna brand was replaced by Ashland in the early 1960s.

****FRONTIER REFINING - BUFFALO, NY:** Founded in Buffalo in 1931, Frontier marketed gasoline in Western New York and Pennsylvania. Frontier merged with Ashland in April, 1950 and Frontier stations were rebranded Ashland about 1960.

****FREEDOM-VALVOLINE - FREEDOM, PA:** Valvoline, founded as the Continuous Oil Refining Company in Binghampton, NY in 1866, introduced the famous trademark "Valvoline" in 1873. So well known were the company's Valvoline lubricants that the company name changed to Valvoline Oil Co. in 1902. In the early 1920s Valvoline began marketing gasoline in New York and Pennsylvania and continued to do so until the 1944 merger with Freedom Oil Co. of Freedom, PA. Freedom had also marketed gasoline since the 1920s and the few remaining Valvoline stations were rebranded Freedom after WWII. Freedom was in the process of discontinuing gasoline marketing, selling its stations to Esso, when the company merged with Ashland Oil in early 1950. Since that time the primary product has been Valvoline Oils, marketed by all the Ashland retail network and throughout the world.

****PAYLESS - NEW ALBANY, IN:** Payless is the postwar creation of James Thornton, operating discount service stations "from Canada to the Gulf Coast" as their slogan claimed. In the middle 1960s Payless acquired the marketing of Louisville's Dance Oil Company. Ashland Oil, long a supplier to both Dance and Payless, bought the operation in about 1970. A few Payless outlets remain today, although most have been converted to SuperAmerica. James Thornton formed Louisville based Thornton Oil Corporation after selling Payless, and continues to operate Thornton stations in six states today.

****SUPERAMERICA - MINNEAPOLIS, MN:** Northwestern Refining Co., the St. Paul Park refining unit of the Erickson family, began operating the SuperAmerica gasoline/convenience store units in about 1962. In 1970 SuperAmerica, a marginal convenience store operation was sold to Ashland Oil and by the middle 1970s began to expand out of it's north central base. Many of the Ashland affiliate brand locations, closed during the gasoline shortages of 1973 and 1978, were rebuilt as SuperAmerica convenience stores and the operation today is considered the model of efficient and customer service oriented gasoline c-store chains.

****OSKEY BROS. PETROLEUM - ST. PAUL, MN:** Oskey Bros. has operated the North Star brand stations in the north central states since the 1920s. Ashland acquired the chain in about 1980 and has since closed many locations or converted them to SuperAmerica. The company has no connection with Canada's North Star Oil Company.

****RED HEAD OIL CO. - WOOSTER, OH:** Red Head, operating discount gasoline stations with service bays, unique in the industry, was founded in Wooster, Ohio in the mid-1930s and operated stations in Ohio, Pennsylvania and West Virginia when they were bought out by Ashland in about 1970. Many have since been converted to SuperAmerica, but some Red Head stations remain today.

****SOUTHERN OIL OF NY - HORSEHEADS, NY:** Southern Oil operated the Rotary brand stations in New York and Pennsylvania from the mid-1920s until the 1970s when the operation was acquired by Ashland Oil.

****HOOD OIL CO. - NORCROSS, GA:** Hood's was a postwar southern operation, rebranded Fleet in about 1960 and acquired by Ashland in about 1967. In the 1980s Fleet was sold to the son of the founder and continues in operation today.

****TRESLER OIL - CINCINNATI, OH:** Tresler was founded in Cincinnati in 1926 and operated retail service stations in Ohio, Indiana and Kentucky until 1985 when they were purchased by Ashland. All have since been converted to SuperAmerica.

****CINCINNATI OIL WORKS COMPANY - CINCINNATI, OH:** Founded in 1841 as a lubricants manufacturer, they marketed gasoline under the White Rose brand (in competition with National White Rose) in a small number of stations. They were purchased by Tresler in 1956.

****HI-FY STATIONS - COLUMBUS, OH:** Discounter purchased by Ashland in the 1970s.

****SAVE MORE and SAVE MART:** Ashland secondary convenience store operations of 1980s and 1990s.

****RICH OIL COMPANY - IRONTON, OH:** Terminal company founded in 1947 that began marketing under the Rich brand in the 1960s. Continues today as an Ashland secondary brand.

Ashland

(GREEN)

1923-1930s | 1930s-1946 | 1946-1954 | 1954-1960

TRI-STATE OIL CO. KENOVA, WV

1930s-1961

ASHLAND SECONDARY BRAND

1946-1960

AETNA OIL AND REFINING COMPANY - LOUISVILLE, KY

1926-1935 | 1930-1938 | 1938-1952 | 1940-1960

1960-TRANSITIONAL

FREEDOM OIL WORKS FREEDOM, PENNSYLVANIA

PRIOR TO 1925 | 1925-1935 | 1935-1940 | 1940-1949

Ashland

(BLUE)
1960-1968

1968-1996

1996-PRESENT

NATIONAL REFINING COMPANY - CLEVELAND, OHIO

1880s-1918

1918-1925

1918-1936

1936-1957

VALVOLINE OIL CORP.
BINGHAMTON, N.Y.

PRIOR TO 1935

1935-1950

c1900-1946

1946-1960

1960-1982

1970s

1982-PRESENT

Ashland

ALLIED OIL, BUFFALO, NY

1950-1960

FRONTIER REFINING COMPANY, BUFFALO, NEW YORK

1931-1955

1955-1960

NORTHWESTERN REFINING CO., SOUTH ST. PAUL, MINNESOTA
SUPER AMERICA, LEXINGTON, KENTUCKY

1962-1985

1985-1994

1994-PRESENT

RED HEAD OIL CO., WOOSTER, OH

1930s

1930s

1930s

1930s-1955

1950s-1970s

c1970-1993

PAYLESS OIL CO. NEW ALBANY, IN

1940s-1960

1960-1971

1965-1970s

1970s-1990s

DANCE OIL SERVICE, NEW ALBANY, IN

1940s

1950s

1960s

OSKEY BROTHERS PETROLEUM
ST. PAUL, MINNESOTA

1920s-1930s

1920s-c1960

c1960-1965

1965-PRESENT

Ashland

MARKETING AREAS TODAY

BI-LO GASOLINE STATIONS, LAWRENCEBURG, INDIANA

BI LO STATIONS INC

1950s-1960s

BI LO

1960s-1970s

BI-LO

1970s-1980s

TRESLER OIL CO., CINCINNATI, OHIO

TRESLER COMET GASOLINE

1926-1938

COMET

1938-1949

COMET

1949-1952

TRESLER COMET

1952-1964

TRESLER COMET

1964-1985

CINCINNATI OIL WORKS, INC. CINCINNATI, OHIO

DEPENDABLE WHITE ROSE GASOLINE

1920s-1956

Ashland

SOUTHERN OIL CORPORATION OF N. Y.
HORSEHEADS, NEW YORK

ROTARY	ROTARY	ROTARY	ROTARY
1930s	1950s-1972	1972-1980s	1980s

HI-FY GASOLINE STATIONS, COLUMBUS, OHIO

HI-FY	HI-FY	HI-FY
1964-1968	1968-1975	1975-1985

HOODS OIL CO., CHAMBLEE, GEORGIA

HOODS	FLEET	FLEET
1950s-1970	1970s	POST 1982

RICH OIL CO., IRONTON, OHIO

RICH	RICH
1947-1984	1990s

ASHLAND SECONDARY BRANDS

Save More	Save Mart

1980s-PRESENT

AtlanticRichfield

THE ATLANTIC REFINING COMPANY - PHILADELPHIA, PA

1915-1933 | 1933-1966 | 1966-1970

RICHFIELD OIL CORPORATION - LOS ANGELES, CA

1915-1926 | 1926-1937 | 1937-1960 | 1960-1966

1966-1970

ATLANTIC RICHFIELD COMPANY - LOS ANGELES, CA

1970-PRESENT

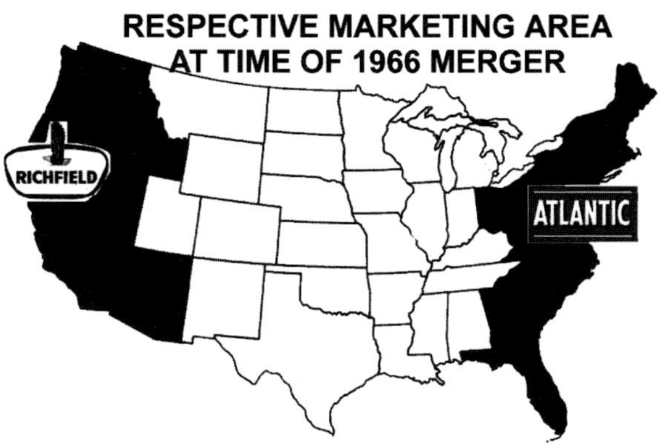

****THE ATLANTIC REFINING CO. - PHILADELPHIA, PA:** The Atlantic Refining Company was founded in Philadelphia in 1866 from earlier interests of the various founders and brought together as a production and refining company. In 1874 Atlantic sold out to Standard Oil and until the breakup of Standard in 1911, functioned as a refining division. Atlantic began marketing in 1915, after the breakup of Standard, and throughout most of its history was in constant search for crude oil to keep its refineries and service stations operating. In 1965 Atlantic, then operating service stations all along the Eastern Seaboard began merger talks with Richfield Oil Corporation of California, a west coast refiner/marketer that traces its roots back to the turn of the century. Atlantic merged with Richfield in 1966, and from 1966 until 1970 marketed under the Atlantic name in the east and Richfield in the west.

****RICHFIELD OIL CORPORATION - LOS ANGELES, CA:** Richfield Oil Corporation was a west coast producer, refiner, and marketer that was founded in California in about 1901. By 1915 Richfield was marketing on the

AtlanticRichfield

RIO GRANDE OIL COMPANY - LOS ANGELES, CA

1930s **1940s** **1950s**

ATLANTIC REFINING - PHILADELPHIA, PA

1985-1991 **1991-1995**

SINCLAIR OIL COMPANY - SALT LAKE CITY, UT

PRIOR TO 1970 PURCHASE **1970-1973** **RESUMED USE AFTER 1973**

SEE ALSO SINCLAIR SECTION IN THIS BOOK

west coast. The company entered east coast markets in the mid-1920s, and created a subsidiary, Richfield Oil Co. of New York with the 1929 purchase of Walburn Petroleum Company. During the 1930s, unchecked expansion met with the Great Depression, and by 1931, Richfield was bankrupt. Sinclair purchased Richfield Oil Company of New York outright in 1935, and in 1936 Sinclair, using assets acquired with Rio Grande Oil Company (purchased several years earlier) participated in the reorganization of Richfield. A new Richfield Oil Corporation was formed in 1937, with Sinclair and Cities Service as owners. With the creation of this entity, Richfield Oil Corporation again became a thriving operation, branding stations under the Richfield and Rio Grande names throughout the west.

****ATLANTIC RICHFIELD COMPANY - LOS ANGELES, CA.**

Formed with the 1966 merger of Atlantic and Richfield, the newly formed operation purchased Sinclair Oil Corporation in 1970. The brandname Arco was introduced at that time to replace all three existing brands, but the Justice Department intervened and required the sale of major portions of Sinclair marketing. From the sale, a new Sinclair Oil Company was created (see Sinclair in this book). The Arco brandname survived on both coasts until 1985, when Arco abandoned marketing in the east, where the name Atlantic was reintroduced by the independent marketing company that had purchased the former Arco marketing. In 1988 the new Atlantic was sold to Sunoco. Despite an effort to rebrand to Sunoco, some stations still operate under the Atlantic name in the northeast today.

****SINCLAIR:** See Sinclair section in this book.

champlin

CHAMPLIN OIL & REFINING COMPANY, ENID, OK

1920-1946 1946-1956 1956-1961

1961-1970

1970-1984

EASON OIL CO., ENID, OK

1920s-1950s

HARBOR STATIONS, ENID, OK

CHAMPLIN SECONDARY BRAND

1960s

EXTENT OF CHAMPLIN MARKETING

****CHAMPLIN OIL AND REFINING - ENID, OKLAHOMA:** Champlin Oil and Refining was an Enid Oklahoma based refiner/marketer founded in 1920. Marketing was concentrated throughout the mid-continent, eventually expanding to 15 states from Tennessee to Colorado. Champlin had long been involved in unbranded marketing and with the purchase of Champlin by the government owned oil company from Venezuela, the decision was made to withdraw from branded marketing. Retail marketing was sold off to American Petrofina and the Fina brand replaced Champlin in 1984. The production and refining divisions survive as an Irving, Texas based subsidiary of Citgo Petroleum which is also owned by the Venezuelan government.

****HARBOR - TULSA, OK:** Harbor is a Tulsa based Champlin secondary brand from the 1960s.

****EASON OIL CO. - ENID, OK:** Eason Oil Co. was an Oklahoma based refiner marketer that was one of the early participants in the Sovereign Service marketing program. By the early 1950s they were based in Enid, OK and eventually merged with Champlin in the mid 1950s.

****CHEVRON AND AFFILIATES STANDARD OIL CO. OF CALIFORNIA:** Standard Oil of California, known in the industry as "Socal", is the west's largest gasoline marketer, tracing it's origins to the 1879 founding of the Pacific Coast Oil Company. Standard Oil, which had maintained a San Francisco marketing office since 1879 purchased Pacific Coast Oil Co. in 1900 and reorganized it as Standard Oil of California.

Standard of California is credited with one of the first retail gasoline stations, opened at the company's bulk plant in Seattle, Washington in 1907. Always an innovator, Standard was soon operating hundreds of retail stations throughout the west. By the mid 1930s, company stations were under the brand "Standard Stations, Inc." and dealer operated locations were "Standard Oil Dealers". In 1945 the Chevron name was introduced for Standard gasolines and independent dealers in the Standard western territory. Dealers in the Standard Texas southern territory and Calso territory became "Chevron Gas Stations", while company owned and operated stations remained "Standard" stations, selling Chevron and Chevron Supreme gasolines. In 1961 Standard of California bought Standard of Kentucky and from 1961 on Chevron gasolines were sold at Standard Oil stations in the southeast. In about 1977 the Standard name came down from all but a very small number of service stations and was replaced by the Chevron name. Standard of Kentucky globes can be found under "Standard Oil" in this book.

****STANDARD STATIONS, INC. AND STANDARD OIL/CHEVRON DEALERS:** Standard Oil Stations was a chain of company owned and operated stations. Standard and Chevron dealers where the dealer operated marketing began Socal in the middle 1930s.

****THE CALIFORNIA COMPANY - ROCKY MTN. MARKETING**

In the late 1930s Standard of California expanded its marketing into the Rocky Mountain area. Not being able to use the Standard name in this area due to its use by Standard of Indiana, gasoline was marketed under the "Calso" (CALifornia Standard Oil) name. In 1946 the Calso stations, all dealer/jobber operated, were rebranded with the new "Chevron Gas Station" logo. Chevron globes, listed above, were used at all Standard Stations in the west and Chevron stations in the Rocky Mountain area.

****CALIFORNIA OIL COMPANY - NORTHEAST MARKETING**

In 1945-46 Standard of California purchased Monogram Oil Co. and set up "The California Oil Company" in Perth Amboy, NJ as an East Coast marketing subsidiary. Stations were branded Calso, same as in the Rockies, and were found from Maine to Virginia. In 1959 Calso Stations were rebranded Chevron.

****PDQ:** Chevron secondary brand in the postwar era.

****SIGNAL OIL COMPANY:** Founded in 1932, Signal Oil sold off their brandname and marketing to Chevron in 1948, before re-entering gasoline marketing in the 1950s under other brandnames (see Signal section). Signal continued as a Socal secondary brand until 1967, when the entire operation was sold to Exxon and rebranded Enco.

****GULF OIL COMPANY:** Purchased by Socal in 1984, stations were either sold off or converted to the Chevron brand. See Gulf section for more information.

****STANDARD OIL "Incorporated in Kentucky" - LOUISVILLE, KY:** Standard Oil of Kentucky was, for 50 years, the world's largest gasoline "jobber". One of the Standard divisions spun off with the 1911 break up of Standard Oil, Standard of Kentucky (commonly called Kyso) was assigned marketing in Kentucky, Florida, Georgia, Alabama and Mississippi, but had no oil refineries, oil production or exploration, or oil pipeline connections. It was simply a marketing group with retail service stations in five southeast states. As a result of this, Kyso contracted with Standard of New Jersey (Esso/Exxon) to supply their needs and simply sold Esso and Esso Extra gasoline under the Crown and Crown Extra names until 1961. When Esso introduced the super-premium "Golden Esso Extra" in 1956, Kyso followed with "Super Crown Extra". This neat arrangement ended in 1961 when Standard of California (Chevron) purchased Kyso outright and cancelled the long-standing Esso supply agreements. Esso scrambled to establish markets in the Kyso territory and Chevron promptly sued them to limit the use of the Standard trademark "Esso" in the five Kyso states. Although Esso had unsuccessfully tried to market under the Esso brandname in Indiana Standard territory (St. Louis 1935-1938), they had long since began using alternate brandnames in other territories. The southeast, however, was different with their already established brand (Kyso sold Esso motor oils) and a court battle ensued. In 1967 Esso lost the rights to the trademark in Kyso territory and service station signs were covered with tiger decals until they could be replaced by Enco signs. After a Supreme Court ruling in 1969 confirming this limitation, Esso began the name search process that eventually led to the Exxon brandname. (See Esso/Exxon section in this book for more on this.) In the meantime, with the 1961 purchase by Chevron, stations were re-imaged and globes were eliminated. The Chevron name replaced "STANDARD" at most Kyso stations in 1977, although a few stations remain Standard to protect the long held brand.

****REFINERS OIL COMPANY - DAYTON, OH:** When Sohio purchased refiners in 1930, Ohio stations were converted to the Sohio brand, and the Indiana and Kentucky operations and the Refiners brandname were sold off to Kyso, which operated it as a secondary brand for several years.

Chevron

STANDARD OIL COMPANY OF CALIF.
SAN FRANCISCO, CA

1915-1931

1930-1946

1938-1957

1940-1942

THE CALIFORNIA COMPANY
NEW ORLEANS, LA

1934-1940

1936-1938

1938-1940

1940-1941

STANDARD OIL DEALERS
CHEVRON DEALERS

1936-1946

1946-1957

1955-1957

1955-1957

CALIFORNIA OIL COMPANY
PERTH AMBOY, NJ
FORMED IN 1945

1945-1949

1949-1959

1959-1970

1957-1970

1970-1977

1970-PRESENT

1940-1942

1942-1946

PDQ STATIONS

1950s

STANDARD OIL COMPANY OF TEXAS EL PASO, TX

1930s

1957-1970

1970-PRESENT

GULF OIL CORPORATION (PURCHASED 1984)

1963-PRESENT

SIGNAL OIL COMPANY, LOS ANGELES, CA

1932-1962

1962-1963 EXPERIMENTAL

1963-1967

LUBRICANTS LOGO

1936-1946

1946-1957

1957-1970

STANDARD OIL COMPANY OF KENTUCKY
LOUISVILLE, KY

1911-1919

1919-1928

1928-1945

1945-1948

1948-1962

1962-1970

1970-1977

1977-PRESENT

REFINERS OIL COMPANY OF KENTUCKY

1930-1933

CITIES SERVICE OIL COMPANY - TULSA, OK: Cities Service was founded in 1910 as a public utility, operating municipal natural gas, lighting, ice, and other utility type services, hence the name "Cities Service". Venturing into petroleum marketing in about 1914, Cities Service was a scattered collection of refining and marketing companies operating gas stations throughout the east and Midwest. By the mid-1930s the various concerns were all marketing under the Cities Service brandname and black and white colors which gave way to green and white stations in 1946. In the 1950s many elaborate service stations were built but Cities Service continued to have an image problem. In 1965 the time-tested Cities Service brandname was replaced by Citgo and the entire marketing program was revitalized. Many new stations were added, only to close in record numbers during the 1970s gas shortages. In 1981 Southland Corporation, operator of the 7-Eleven convenience stores, and the government owned oil company in Venezuela purchased the marketing division, Citgo Petroleum, while Occidental Petroleum purchased the remaining divisions of Cities Service Oil Company. In the late 1980s, Southland, in bankruptcy, sold their share of Citgo to Venezuela and the company, with the assured supply of a government owned oil company, again began aggressively marketing. Citgo now supplies thousands of stations in 46 states and new locations are added every day.

****LOUISIANA OIL REFINING CO. - SHREVEPORT, LA:** Originally a Gulf Coast refiner/marketer, Loreco became a part of Cities Service in 1932.

****CREW-LEVICK COMPANY - PHILADELPHIA, PA:** Crew-Levick was an early Philadelphia lubricants marketer that merged with Cities Service in about 1915.

****WARNER-QUINLAN COMPANY - NEW YORK, NY:** Founded in New York City before 1900, Warner-Quinlan was marketing throughout New England when purchased by Cities Service in 1933.

****SHOEMAKER OIL CO - AUBURN, NY:** Purchased by Cities Service in about 1935. Little is known about this company.

****JENNEY MANUFACTURING CO. - BOSTON, MA:** Originally a mid-19th century dealer in whale oils, coal, kerosene and related products, Jenney entered the New England gasoline market in the early years of this century. By 1920 Jenney had a network of elaborate service stations in Boston and the surrounding area. With close ties to Cities Service for many years, Jenney was merged into Cities Service in about 1965 and the Citgo brand replaced Jenney.

****LINDSAY-McMILLAN OIL COMPANY - MILWAUKEE, WI:** Independent marketer purchased by Cities Service in the late 1920s.

****A. R. NEWCOMB OIL COMPANY - KINGSTON, NY:** Independent marketer purchased by Cities Service in the late 1920s, they operated under the Keystone brandname.

****DIXCEL:** Missouri brand originally operated by Pioneer Oil Company and later Milton Oil Company, Cities Service purchased the Dixcel operation in 1962.

****THE SOUTHLAND CORPORATION - DALLAS, TX:** Originally a municipal ice distributor, Southland is credited with inventing the convenience store when in 1927 a Dallas ice dock operator added a stock of staple items to his wares. Stores were originally known as "Totem Stores", but in 1946 the Seven-Eleven brandname was introduced. Gasoline tie-ins (with Transcontinental Oil) were known before 1930, but the "Seven-Eleven" branded gasoline operations were not formally established until about 1970. When Southland purchased a portion of Cities Service in 1981, virtually all of the Seven Eleven stores converted their gasoline operations to Citgo.

TRADITIONAL CITIES SERVICE MARKETING AREA

CITIES SERVICE

<u>CITIES SERVICE OIL COMPANY</u>

1914-1932 | 1932-1934 | 1932-1946 | 1946-1955

<u>LINDSAY-McMILLAN CO</u>
<u>MILWAUKEE, WI</u>

1920s

<u>A.R. NEWCOMBE OIL CO</u>

1920s

<u>LOUISIANA OIL</u>
<u>REFINING CO</u>
<u>SHREVEPORT, LA</u>

PRE-1934

<u>JENNEY MANUFACTURING COMPANY</u>
<u>BOSTON, MA</u>

1912-1920s | 1920s-1940 | 1940-1955 | 1955-1966

CITIES SERVICE

CITIES SERVICE OIL COMPANY / CITGO PETROLEUM

1955-1957

1957-1960

1960-1965

1965-PRESENT

CREW LEVICK AND COMPANY
PHILADELPHIA, PA

1915-1926

1926-1934

WARNER-QUINLAN
OIL COMPANY
NEW YORK, NY

PRE-1933

MILTON OIL COMPANY
SEDALIA, MO

1930s

1940s-1950s

THE SOUTHLAND CORPORATION
DALLAS, TX

1970s-1982

CLARK OIL AND REFINING COMPANY, MILWAUKEE, WI

1932-1938 1938-1946 1946-1954

1954-1994 1994-PRESENT

OWENS OIL COMPANY BLOOMINGTON, IL

1930s-1960 1960-1970s

TARGET STATIONS, MILWAUKEE, WI

1960s

****CLARK OIL AND REFINING CORP. - ST. LOUIS, MO:** Clark Oil Co. began as a one station operation by Mr. Emory Clark in 1932 in Milwaukee, Wisconsin. By the early 1930s, Clark had evolved into a chain operation, selling "discount" gas through company operated stations in a number of Wisconsin towns. Significant growth came after World War II, when the operation expanded to include stations from Pennsylvania to Kansas, totalling more than 1350 at the peak, including 122 stations acquired when Clark bought out the Owens Oil Co. of Bloomington, IL in 1968. In 1981 Clark was purchased by St. Louis based Apex Oil and in 1988 again sold, this time to Canadian conglomerate Horsham Corp., which continues to operate the more than 900 Clark stations in 12 states today.

****TARGET:** Target is a 1960s Clark secondary brand**

****OWENS OIL CO. - BLOOMINGTON, IL:** Owens was a Bloomington, IL based discounter purchased by Clark in 1968. Use of the Owens brandname continued for many years after the merger.

DERBY OIL COMPANY, WICHITA, KS

1920-c1930

1929-1949

1950-1988

SOVEREIGN SERVICE

1937-1955

COASTAL OIL AND REFINING

1988-PRESENT

1988-PRESENT

****COASTAL CORPORATION - HOUSTON, TX**

****DERBY OIL CO: WICHITA, KS:** Derby Oil Co. was a mid-continent refiner/marketer founded in Wichita, Kansas in 1920. In the late 1930s, Derby was instrumental in the creation of Sovereign Service, a marketing organization for independent dealers supported by a number of mid-continent refiners. After WWII, Derby concentrated on an expanding chain of retail service stations as well as numerous dealer operated units and a large jobber/marketer arrangement with Piester. In 1955 Colorado Interstate Gas Company purchased Derby and with its large oil production system allowed for greater expansion of the Derby refinery and marketing network. In the early 1970s Colorado Interstate Gas became a part of the Coastal Corporation and in 1988 Derby stations, along with others in the Coastal organization were rebranded Coastal.

****SOVEREIGN SERVICE:** Sovereign Service was a franchise program allowing independent dealers supplied by the various mid-continent refiners that participated in the program to be branded "Sovereign Service". Dealers were held to high standards of operation, supported by co-operative advertising of the programs' features (independent dealers only, clean rest rooms, 24 hour service, etc.) Derby was a leading participant in the program, along with Kanotex/Apco, Vickers, Rock Island, Eason, Elreco, Yale, Shallow Water and perhaps others. The program was initiated in the late 1930s (about 1937) and survived as late as 1959. Globes from the individual brands are listed with that company's listing in this book.

CONTINENTAL OIL COMPANY, PONCA CITY, OK

1913-1929

1929-1950

1950-1970

1970-PRESENT

MARLAND OIL COMPANY PONCA CITY, OKLAHOMA

1920-1929

NATIONAL OIL COMPANY RICHMOND, VA

PRE-1937

F. P. KENDALL OIL COMPANY, CHATTANOOGA, TN

1940-1959

1959-1980

1980-1992

1980-PRESENT

DOUGLAS OIL COMPANY, LOS ANGELES, CA

1928-1940s

1930s-1940s

1946-1970

1965-1971

KRIEGER OIL COMPANY

PRIOR TO 1941

WESTERN OIL AND FUEL COMPANY, MINNEAPOLIS, MN

1935-1946 | 1946-1955 | 1955-1960

ZENITH GAS POOL
DULUTH, MN

1930s | 1950-1960 | 1960-1971 | 1968-1971

DIRECT SERVICE, MINNEAPOLIS, MN

1930s-1950 | 1950-1960 | 1960-1971

WESTERN
BRAND INTRODUCED 1958

1958-1960 | 1960-1969 | 1969-1980

MALCO OIL COMPANY
ROSWELL, NM

1930s-1950 | 1950s

VALLEY AND NUMEX
MALCO SECONDARY BRANDS

1950s | 1950s

CONOCO SECONDARY BRANDS

E=QUAL — 1970s

FAS GAS — 1970s

CONTINENTAL OIL COMPANY - PONCA CITY, OK: Continental Oil, founded in 1875 in Denver, Colorado became an operating unit of the Standard Oil Trust, having been purchased by Standard in 1885. After the 1911 breakup of Standard, Continental was spun off as an independent company and by 1913 was involved in production, refining and marketing. The Continental name survived a 1929 takeover by Marland Oil, and over the course of time the Conoco brandname has appeared in nearly all states of the Southeast, Midwest, Mountain and Pacific Coast areas. Conoco expanded it's scope worldwide during the 1950s, including the purchase of several marketing companies, such as Western Oil, Kayo and Douglas, greatly expanding the company's marketing presence throughout the United States. Conoco has been a part of the giant chemical company Dupont since 1982 and markets in 37 states today.

MARLAND OIL COMPANY - PONCA CITY, OK: Marland traces its early history to the Oklahoma oil fields of the early 1910s. By 1920 marketing began under the famous Red Triangle logo and by 1922 there were Marland service stations in 11 states. In 1929 Marland purchased Continental but chose to retain the Continental Oil Co. name for the new corporation. The new Conoco logo was to be the famous Marland triangle rebranded with the name "CONOCO".

WESTERN OIL AND FUEL - MINNEAPOLIS, MN: Western Oil and Fuel was a Minneapolis based marketing operation, founded in 1926, that introduced the famous Husky brand in the mid 1930s and operated the Husky stations into the late 1950s. They had also purchased a marketing concern, Zenith Gas Pool, and adopted their "Mileage" brandname. Conoco bought out Western in 1958 and at about that time introduced "Western" as another gasoline trademark. By 1970 Conoco had converted all the Mileage and DS locations to the Conoco brand, but the Western stations, modeled after the competing Holiday and SuperAmerica convenience stores, remained "Western" up through the late 1970s when they were rebranded "JET".

DIRECT SERVICE OIL CO. - MINNEAPOLIS, MN: Minneapolis marketer purchased by Western Oil and Fuel in 1956.

MALCO REFINING - ROSWELL, NEW MEXICO

Malco was a Roswell, New Mexico refiner marketer operating stations in New Mexico under the Malco and Numex brands and under the Valley brand in Arizona. Conoco purchased Malco in 1959 but was forced to sell on anti-trust grounds in 1964. Malco operated as an independent into the 1970s.

VALLEY: Malco secondary brand from the 1950s.

NUMEX: Malco secondary brand from the 1950s.

DOUGLAS OIL CO. - LOS ANGELES, CA

Douglas Oil dates back to about 1927 and merged with Krieger in 1946. In 1961 Conoco bought Douglas and operated the nearly 300 Douglas stations until 1971 when they were sold off and rebranded, mostly as Texaco.

KRIEGER: California independent that merged with Douglas in 1946.

F. P. KENDALL AND CO. - CHATTANOOGA, TN: F.P. Kendall and company opened their first Kayo station in Chattanooga, Tennessee in 1940. The chain expanded throughout the south and had approximately 170 stations in seven states when purchased by Conoco in 1959. Kayo was operated as a separate brand throughout the 1960s and 70s and in about 1980 the nearly 900 Kayo stations rebranded "JET". Conoco began replacing the Jet brand in 1990.

FAS-GAS and E=QUAL: Conoco secondary brands of the 1970s.

****FARMLAND INDUSTRIES - KANSAS CITY, MO:** Founded in 1928 as Consumers Co-operative Association, Farmland, commonly referred to as "Double Circle" in reference to their logo, supplies products to farm co-ops throughout the Midwest. They continue to market through co-ops under the "CO-OP" brandname and at retail under the "AMPRIDE" brand.

****FALCON:** A postwar attempt at retail markets by Consumers Co-Operative Association (CCA).

****FARMERS UNION CENTRAL EXCHANGE - ST. PAUL, MN:** Farmers Union Central Exchange is a combination of several farm co-operative organizations that operate across the North Central states. The CENEX brandname was introduced in about 1974 to establish a common identity for the various companies. In 1988 Farmers Union Central Exchange merged with Land O'Lakes and all petroleum products are marketed under the CENEX brandname today.

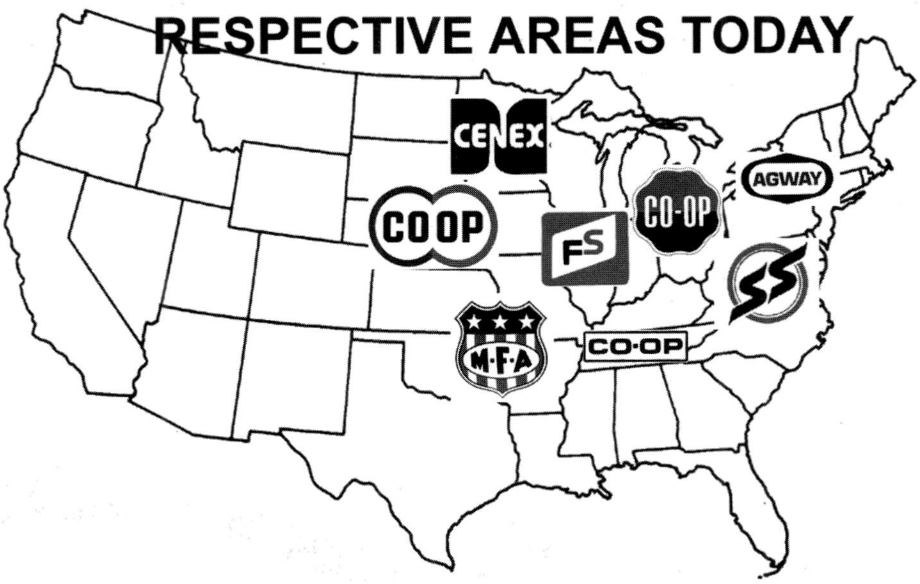

****MIDLAND COOPERATIVES - ST. PAUL, MN:** Midland, founded in 1926, sold out to Land O'Lakes in 1985 and Land O'Lakes merged with Farmers Union Central Exchange (CENEX) in 1988. All gasoline has been sold under the Cenex brand since that time.

****FARMERS EDUCATIONAL AND CO-OPERATIVE STATE UNION - OMAHA, NE:** Nebraska farmers' co-operative that was purchased by Farmers Union Central Exchange in the 1960s.

****SOUTHERN STATES - RICHMOND, VA:** Founded as Virginia Seed Service in 1923, Southern States began marketing petroleum products in the 1930s and continues today.

****TENNESSEE CO-OPERATIVES- LaVERGNE, TN:** Tennessee's farmers co-operative, continuing in operation today.

****GROWMARK INDUSTRIES - BLOOMINGTON, IL:** Founded as Illinois Farm Service, they marketed gasoline retail under the "Magic Aladdin" brandname. Continues to operate today under the FS brand.

****MISSOURI FARMERS ASSOCIATION - COLUMBIA, MO:** MFA has marketed petroleum products since the 1930s, and continues to market through farm co-operatives and some retail locations in Missouri and Western Tennessee.

****CENTRAL CO-OPERATIVE WHOLESALE - SUPERIOR, WISCONSIN:** Farm co-operative that operated in Wisconsin and Minnesota.

****CO-OPERATIVE GLF (Grange League Farm) - ITHACA, NY:** Founded in the 1920s, GLF served farmers in New York and Pennsylvania as well as parts of New England. In 1964, GLF merged with Eastern States Farm Supply to form Agway and continues to market petroleum products under that name throughout the northeast.

****LANDMARK CO-OPERATIVES - COLUMBUS, OHIO:** Marketed in Ohio under the Landmark and Fyr-Zon brandnames.

****FARM BUREAU CO-OPERATIVE - INDIANAPOLIS, IN:** Indiana's farmers co-operative, they continue to market under the Farm Bureau brand in Indiana today.

CONSUMERS CO-OPERATIVE ASSOCIATION, KANSAS CITY, KS

1928-1931 | 1931-1937 | 1937-1950 | 1950-1958

FALCON STATIONS, INC

1950s

AMPRIDE

1990-PRESENT

1958-PRESENT

TENNESSEE FARMERS CO-OPERATIVES, LaVERGNE, TN

1930s-1950s | 1950s-1970 | 1970-PRESENT

CENTRAL CO-OP SUPERIOR, WI

1930s

UNAFFILIATED CO-OPERATIVES

1950s

UNITED FARMERS OF ALBERTA, EDMONTON, AB

1930s-1960s | 1960s-PRESENT

FARMERS UNION CENTRAL EXCHANGE
INVER GROVE HEIGHTS, MN

1930s

1960-1968 | 1946-1968 | 1968-1973 | 1973-PRESENT

MIDLAND CO-OPERATIVES / LAND O'LAKES CO-OPERATIVES
ST. PAUL, MN

1930-1946 | 1946-1959 | 1959-1985

1959-1985 | 1985-PRESENT | 1989-PRESENT

FARMERS UNION STATE EXCHANGE, OMAHA, NEBRASKA

1940s | 1940s-1958 | 1958-1961 | 1961-1970s

SOUTHERN STATES CO-OPERATIVES, RICHMOND, VA

1934-1962

1962-1985

1985-1992

1992-PRESENT

FARMERS CENTRAL EXCHANGE, RALEIGH, NC

1930s-1970

1946-1970

1970-1982

MISSOURI FARMERS ASSOCIATION, COLUMBIA, MO

1970-PRESENT

1980s-PRESENT

1930s-1950

1950-1970

ILLINOIS FARM SERVICE / GROWMARK INDUSTRIES, BLOOMINGTON, IL

1930s-1947 **1947-1960** **1960-PRESENT**

1930s-PRESENT **1950s**

GRANGE LEAGUE FARM CO-OPERATIVES, ITHACA, NY

1930s-1940 **1940-1965** **1965-PRESENT**

FARM BUREAU SERVICES, INC., INDIANAPOLIS, IN

1930s-1955 **1955-1982** **1982-PRESENT**

CROWN CENTRAL PETROLEUM CORPORATION, BALTIMORE, MD

c-1920-c1930

1930-1938

1938-1945
LIMITED USE THRU 1956

1945-1966

1966-1973

1973-PRESENT

ENERGY OIL COMPANY ANNANDALE, VA

c1950-1967

1967-1973 & 1984-1987

THRIFT OIL

1950s SECONDARY BRAND

FAST FARE STORES, HENDERSON, NC

1985-PRESENT

1985-PRESENT

****CROWN CENTRAL PETROLEUM - BALTIMORE, MD:** Crown Central Petroleum traces its earliest history to the east Texas oil fields where in 1917 oil exploration got underway. By 1900, a Houston refinery was in operation providing high quality lubricants. In 1930, Crown's headquarters shifted to Baltimore, Maryland under new ownership and in 1933 retail marketing in the east was added to a chain of service stations already operated in Houston. Throughout the 1930s and 40s jobbers were actively sought in markets from Pennsylvania to Alabama. In the mid-1960s Crown became involved in direct retail marketing with the addition of the "Peoples" stations of Energy Oil Co. So successful was direct retailing that by the early 1970s jobber accounts were abandoned in favor of all company operated stations. Crown also entered the convenience store market in the early 1980s with the purchase of Sunoco's "Fast Fare" chain in the Southeast. Crown products are currently available at over 600 locations in the Southeast.

****THRIFT:** A 1950s Crown secondary brand.

****ENERGY OIL COMPANY - ANNANDALE, VA:** Founded as Peoples Self-Service Gas Stations, Inc., in Annandale, Virginia in the late 1940s, Peoples operated a chain of retail gasoline stations in Virginia, Maryland and Washington, DC. By the mid-1960s, the corporate name had become "Energy Oil Company" and over 20 stations were being operated. In 1972 Crown, who had supplied Peoples for years, purchased the chain and rebranded the stations Crown. Peoples then became a Crown secondary brand, appearing on a few remaining jobber-operated stations in North Carolina. The name was re-introduced for use on some company operated stations Crown purchased from Ashland in 1985, but these were rebranded Crown in 1987 and the People brand is no longer in use.

CRYSTAL

MERCURY OIL COMPANY, SHREVEPORT, LA

1940s-1961

1961-1968

CRYSTAL OIL AND REFINING COMPANY, SHREVEPORT, LA

1968-1979

1968-1979

TULSA OIL COMPANY, DETROIT, MI

1930s-1946

1946-1985

STONES OIL COMPANY, FITZGERALD, GA

1960s

R. J. OIL REFINING COMPANY TERRE HAUTE, IN

1930s-1971

EASTERN OIL COMPANY MONTGOMERY, AL

1960s

BERRY PETROLEUM, STEPHENS, ARKANSAS

1950s

HI-OCTANE TERMINAL COMPANY, PANAMA CITY, FL

1946-1973

****CRYSTAL OIL COMPANY, SHREVEPORT, LA:** Crystal Oil was formed in 1968 when production company Crystal Oil and Land Company purchased the Mercury Oil Company chain of service stations in Shreveport, LA. Determined to be a large independent refiner-marketer, Crystal then set about to purchase a series of brands. Acquisitions include Tulsa Oil Company of Detroit, an old independent later sold to Rock Island, R. J. Oil Refining of Terre Haute, IN, Eastern Oil Company of Montgomery, AL, Stones Oil Company of Fitzgerald, GA, Berry Petroleum of Stephens, AR, and Hi Octane Terminal Company of Panama City, FL. Crystal struggled through the gas shortages of the 1970s, finally selling off what operations they could and closing the rest in the early 1980s.

DIXIE DISTRIBUTORS, ANN ARBOR, MI

1926-1950s

1950s-1960s

McNUTT OIL AND REFINING COMPANY / EL PASO GAS AND OIL, EL PASO, TX

1954-1959

PRE-1964

DIXIE DISTRIBUTORS, INC. - ANN ARBOR, MI: Dixie Distributors, Inc., was a distributor association put together by a group of Michigan based oil jobbers in Ann Arbor, Michigan in 1926. Almost immediately independent jobbers from all around the country applied for a Dixie franchise and joined the association in order to be able to market their products under the Dixie name. The organization expanded greatly until World War II, when gasoline shortages forced many independents, including many of the Dixie affiliates out of business. Marketing efforts by the majors after WW II were so great that an independent organization could not hope to compete and many former Dixie jobbers eventually were branded by the majors. Nevertheless, Dixie stations survived until the 1960s or later in some areas, and the organization survives today.

ERICKSON

****ERICKSON OIL COMPANY - MINNEAPOLIS, MN:** Founded in Wisconsin in 1928, Erickson expanded from a country store operation into gasoline marketing in 1931. They developed a chain of service stations in the years following World War II, and in 1959 launched the Holiday Stationstores, the first gasoline oriented convenience stores in existence and forerunner to most of today's gasoline marketing operations. Stations operate under the Holiday brand in the north central states today.

****LIFE:** Erickson secondary brand from the early 1950s.

****ERICKSON POST**: Secondary convenience store chain in operation in the 1990s.

****NORTHWESTERN REFINING:** The Erickson family interest led to an investment in a twin cities refinery in the postwar years. Northwestern Refining entered branded gasoline retailing in 1962 with the SuperAmerica convenience store chain. Northwestern was sold to Ashland in 1970, and Ashland continues to operate a greatly expanded SuperAmerica today.

****ERICKSON OIL PRODUCTS - HUDSON, WI:** Erickson is a family owned operation, and in 1950 one of the family members established the competing Erickson Oil Products operation across the river from the original Twin Cities base. Erickson continues to operate under the Erickson and Freedom brandnames today.

ERICKSON

ERICKSON OIL COMPANY, MINNEAPOLIS, MN

1931-1946

1946-1970s

1959-1985

1985-1990

1990-1993

1993-PRESENT

LIFE GASOLINE STATIONS

1950-1954

ERICKSON POST

1990s

NORTHWESTERN REFINING CO ST. PAUL PARK, MN

1962-1985

(SEE ASHLAND)

ERICKSON OIL PRODUCTS, HUDSON, WI

1950-1970s

1970s

1970s

1979-1990

1990-PRESENT

EXXON

STANDARD OIL COMPANY OF NEW JERSEY

"STANDARD"
GASOLINE

1915-1923

1923-1924

1924

1924-1933

1970-1971
EXPERIMENTAL

1972-1982

1982-PRESENT

HUMBLE OIL & REFINING
HOUSTON, TEXAS

1919-1930

1920s

1930-1933

1933-1946

1946-1962

THE POWERINE CO., DENVER, COLORADO

1920s-1938

1938-1946

1946-1950

EXXON

ESSO STANDARD OIL COMPANY

1926-1933 | 1933-1946 | 1933-1940 | 1933-1965 | 1965-1972

TEXAS PACIFIC OIL COMPANY

PRE 1945

STANDARD OIL CO. OF LOUISIANA NEW ORLEANS, LA

1916-1923

COLONIAL BEACON OIL CO., BOSTON, MA

1910s-1930 | 1910s-1931 | 1931-1933 | 1933-1946

NORTHWESTERN OIL CO. CUT BANK, MONTANA

1930s | 1930s-1945

YULE OIL CO. BILLINGS, MT

1930s-1945

CARTER OIL CO. TULSA, OKLAHOMA

1945-1946

SUN-FLASH, COLUMBUS, OHIO

1930s | 1946-1961

HUMBLE OHIO OPERATIONS

1961-1965 | 1965-1972

OKLAHOMA OIL CO.
CHICAGO, IL

1936-1957

1957-1960

PERFECT POWER STATIONS, INC.
CHICAGO, ILLINOIS

1935-1956

GASETERIA-BONDED, INDIANAPOLIS, INDIANA

1946-1956

1930-1946

1950-1956

HOOSIER PETROLEUM, INDIANAPOLIS, INDIANA

1935-1956

1950s

1930-1956

JOHNSON OIL CO.
CHICAGO, ILLINOIS

1920s-1956

1946-1956

BRILLIANT BRONZE
JOHNSON'S SECONDARY BRAND

1940s

1950s

EXXON

ENCO-INTRODUCED 1960

1960-1965

1964-1965

1965-1980

TRANSITIONAL AFFILIATE

1970s

PATE OIL CO., MILWAUKEE, WI

1933-1956

1956-1961

ALERT OIL CO. WILMINGTON, DE

1969-1984

SOUTHERN OIL MARKETING, MIAMI, FL

1950s

IMPERIAL OIL, LIMITED TORONTO, ONTARIO

1880s-1920

1919-1925

1925-1934

1934-1946

1946-1965

1965-PRESENT

EXXON

CHAMPLAIN OIL, LTD.
MONTREAL, QUEBEC

CARRIBBEAN MARKETING

LUBRICANTS MARKETING

CHAMPLAIN — **1940s-1976**

1976-1990s

CREOLE — **1950s**

Penola — **1950s**

HOME OIL COMPANY, VANCOUVER, BRITISH COLUMBIA

HOME GAS — **PRIOR TO 1964**

HOME — **1964-1972**

HOME — **1972-1980s**

EXTENT OF REGIONAL BRANDS, 1959

****EXXON AND AFFILIATES**
****STANDARD OIL COMPANY OF NEW JERSEY - NEW YORK, NY:** Esso is, of course, Standard Oil of New Jersey, largest of the Standard companies to emerge from the 1911 breakup of Standard Oil. Standard (NJ) began marketing in several Eastern Seaboard states and in Tennessee, Louisiana, and Arkansas through Standard of Louisiana, a subsidiary assigned to them with the 1911 breakup. The Esso brand was first introduced in 1926 as an Ethyl grade gasoline and in 1933 replaced Standard as the company's brandname. Esso marketing expanded to include stations in 19 states with marketing after WWII added throughout the country under various brandnames detailed below. In 1959 all domestic marketing was grouped under the subsidiary Humble Oil and Refining, formerly an independent company that Standard purchased in 1919. Under Humble, the Esso brandname was retained in the East and South. Enco was introduced to replace the various other regional brands and Humble was used in Ohio when Sohio objected to both Esso and Enco. In 1972 the brandname Exxon was introduced to replace Esso, Enco and Humble, although Enco remained in use in the upper Midwest area until that area was abandoned several years later. Exxon continues to market through over 11,000 stations in 40 states.

****EXXON:** The Exxon brandname was introduced in the fall of 1972 replacing Esso in 19 eastern states, Humble in Ohio and Enco elsewhere. When Exxon was introduced the Enco brandname remained in use in a limited number of stations in the Chicago marketing area that had already been scheduled for abandonment

in about 1976.
**STANDARD OIL CO. OF LOUISIANA: Marketing affiliate assigned to Standard NJ at the 1911 breakup of Standard Oil. Stanocola operated stations in Tennessee, Arkansas and Louisiana under the Stanocola brand until 1923 when the "STANDARD" bar and circle image was introduced throughout the marketing territory.
**HUMBLE OIL AND REFINING - HOUSTON, TX: Houston based producer/refiner/marketer purchased by Standard (NJ) in 1919. The Humble brand was used in Texas and New Mexico from 1919 until 1961. In 1959 all United States marketing came under the subsidiary "Humble Oil and Refining" and "Humble" was added to stations of each brand then in use, although each retained its own regional brandname (Esso, Carter, Oklahoma, etc.) In 1961 Enco replaced Humble as the station brand in Texas and Oklahoma. The Humble brand was introduced in Ohio when both Esso and Enco were blocked by Sohio and remained in use in Ohio until converted to Exxon in 1972.
**TEXAS PACIFIC COAL AND OIL - FT. WORTH, TX: Independent Texas marketer sold out to and rebranded Humble in 1945.
**COLONIAL BEACON OIL COMPANY - BOSTON, MA: Formed in a late 1920s merger between Boston area based Colonial Oil Company and Beacon Oil. Colonial Beacon, an innovative marketer with stations throughout New England, sold out to Standard (NJ) in 1930. The Colonial name, as a subsidiary, survived until the end of WWII, although products were branded Esso in 1933. Colonial influenced the entire industry with the introduction of service bays at stations and the marketing of tires, batteries and automotive accessories (known as TBA in the trade) through their stations. Colonial was instrumental in the formation of the Atlas Supply Company to supply the various Standard companies that participated in the TBA line program.
**POWERINE OIL CO. - DENVER, CO: Denver based refiner/marketer. Purchased by Standard (NJ) in 1938, the Powerine name was retained until the end of WWII when it was replaced by Oval-E.
**NORTHWEST REFINING CO. - CUT BANK, MT: Marketed Grizzly gasoline. Purchased by Standard in 1942 and rebranded Carter in 1945.
**OVAL E: Name introduced in 1946 to replace Powerine. Changed to Carter in 1950.
**YALE OIL CO. - BILLINGS, MT: Marketed Litening gasoline. Purchased by Standard (NJ) in 1946 and rebranded Carter.
**CARTER OIL CO. - TULSA, OK: Long a Standard production subsidiary, the Carter brandname was introduced in 1945 to replace Grizzly and Litening. In 1950 Carter replaced Oval-E to bring all of Standard's western marketing under the Carter brand. The Carter brand was replaced by Enco in 1961.
**ENCO: Introduced in 1960, first in the Oklahoma/Pate marketing territory as a replacement for those brandnames and expanded to Carter and Humble territories in 1961, Enco (from the slogan America's Leading ENergy COmpany), was to be Standard's primary brandname in all areas where Esso could not be used. Replaced by Exxon in 1972 except in select areas where Standard planned to withdraw from marketing. Used as late as 1980 in those areas.
**PATE OIL CO. - MILWAUKEE, WI: Founded in Milwaukee in 1933, Pate operated over 50 stations when purchased by Standard in 1956. The Pate brand was replaced by Enco in 1960.
**OKLAHOMA OIL COMPANY - CHICAGO, IL: Oklahoma was a Chicago area discount marketer that Standard (NJ) purchased in 1956. Almost immediately another Chicago discounter, Perfect Power, was purchased and rebranded Oklahoma. The Gaseteria/Hoosier Pete operation was added in 1957 and also rebranded Oklahoma. The Enco name was first introduced to replace Oklahoma in 1960.
**GASETERIA OIL CO. - INDIANAPOLIS, IN: Founded in 1928 in Indianapolis, Gaseteria operated the Bonded stations throughout Indiana and Kentucky. They also purchased the remnants of Johnson Oil Refining/Brilliant Bronze, as well as Hoosier Petroleum in 1956. In 1957 Standard purchased Gaseteria and rebranded the operation Oklahoma and the Ohio locations were spun off as an independent company, based in Springfield, Ohio.
**JOHNSON OIL REFINING CO. - CHICAGO, IL: Mid-Continent refiner/marketer that marketed gasoline under the Johnson, Jo-Re-Co, and Brilliant Bronze brandnames. Gaseteria purchased Johnson in 1956 and sold the refinery operation to Deep Rock. Rebranded Oklahoma with the 1957 purchase of Gaseteria by Standard (NJ).
**HOOSIER PETROLEUM CORP. - INDIANAPOLIS, IN: Indiana marketer purchased by Gaseteria in 1956. Founded in 1930, they marketed under the Hoosier Pete and Vortex brandnames.
**RYAN'S: A Hoosier Pete affiliate.
**SOUTHERN OIL CO. - MIAMI, FL: Purchased by Standard (NJ) in 1962 and rebranded Esso.
**SUN-FLASH OIL CO. - COLUMBUS, OH: Purchased by Standard (NJ) in 1961 and rebranded Humble.
**ALERT OIL COMPANY - WILMINGTON, DE: Alert was formed in the late 1960s as an Esso secondary brand. All converted to Exxon in the 1980s.
**IMPERIAL OIL, LTD. - TORONTO, ONTARIO: Founded in 1884, Imperial became the Canadian branch of Rockefeller's Standard. At the breakup of Standard in 1911, control of Imperial was assigned to Jersey Standard, and remains such to this day. Markets across Canada under the Imperial Esso brand.
**CHAMPLAIN OIL PRODUCTS, LTD. - MONTREAL, PQ: Regional marketer acquired by Imperial Oil in the 1960s.
**HOME OIL PRODUCTS, LTD. - VANCOUVER, BC: Western Canada regional marketer, purchased by Imperial Oil in the 1970s.

FINA

PANHANDLE REFINING WICHITA FALLS, TEXAS

1926-1948 | 1948-1956 | 1956-1958

AMERICAN LIBERTY OIL COMPANY DALLAS, TEXAS

1930s-1946 | 1946-1957 | 1957-1958

PETRO ATLAS CORPORATION ELDORADO, KANSAS

1930s | 1930s-1955 | 1955-1957 | 1957-1958

EXTENT OF FINA MARKETING

FINA

AMERICAN PETROFINA, DALLAS, TEXAS

1956-1973

1973-PRESENT

1973-PRESENT

COSDEN OIL and REFINING BIG SPRING, TEXAS

1926-1929

1929-1938

1938-1946

1946-1960

1960-1963

COLONIAL OIL COMPANY DES MOINES, IOWA

1950s

COSDEN-SECONDARY BRANDS

1955-1963

1930s-1963

WIDES OIL CO. DES MOINES, IOWA

1940s-1970s

PETROLEUM MARKETERS TULSA, OK

1950s

LIBERTY PETROLEUM MOUNT VERNON, ILLINOIS

1940s-1966

1966-1970s

FINA

AMERICAN PETROFINA - DALLAS, TX: American Petrofina is the Dallas, Texas based subsidiary of Belgium's PetrofinaSA. American Petrofina has been formed almost entirely by mergers and buyouts, beginning with the purchase of Panhandle Oil Corp. of Wichita Falls, Texas in October 1956. Subsequent purchases, where an entire company's marketing was purchased and converted, are listed below. Fina currently operates stations throughout the Mid-continent and Southeast.

PANHANDLE OIL COMPANY - WICHITA FALLS, TX: Established in 1923, Panhandle marketed in Texas. American Petrofina entered U.S. markets with the purchase of Panhandle Oil in October 1956. The brand was phased out by 1959.

AMERICAN LIBERTY - DALLAS, TX: Formed in the 1930s from oil holdings of the Murchison family, sold out to Fina in October 1957.

PETRO-ATLAS - EL DORADO, KS: Petro Atlas is a consolidation of the old Eldorado Refining and K-T Refining. Petro-Atlas sold out to Fina in March 1958.

PETROLEUM MARKETING CO. - TULSA, OK: Pemco is a Tulsa based independent that rebranded Fina in about 1960.

COLONIAL OIL COMPANY, DES MOINES, IA: Affiliate with Petro-Atlas, the Colonial operations were purchased by American Petrofina in the early 1960s.

COSDEN PETROLEUM - BIG SPRING, TX: Founded in about 1925 by Josh Cosden after he lost controlling interest in Mid-Continent Petroleum which he had founded in 1916. Cosden was operating a Big Spring,TX refinery and petrochemical operation and marketing through stations in six states when they merged with American Petrofina in April 1963. Cosden stations were rebranded Fina shortly after the merger.

COL-TEX: The Col-Tex brand was sold off to Cosden after Apco purchased and rebranded Kanotex, greatly expanding the Apco brand from it's original Oklahoma marketing territory. When Apco became the primary brandname, Col-Tex stations were either converted to Apco or sold off with the Col-Tex brand to Cosden, who operated the Col-Tex stations as a secondary brand until 1963.

ONYX: Was a Texas independent that sold out to Cosden and beacame a Cosden secondary brand.

WIDES OIL COMPANY - MURPHYSBORO, IL: Wides is a Murphysboro, Illinois based independent bought by Fina in about 1972 and converted to the Fina brand after 1977.

LIBERTY OIL COMPANY - MT. VERNON, IL: Regional independent dating back to the 1930s and purchased by Fina in the late 1970s.

Getty

FLYING A/GETTY
TIDEWATER OIL COMPANY - NEW YORK, NY
Founded in New York in 1887, Tidewater is best known for their world famous motor oil brand "Veedol". Tidewater entered the gasoline market about 1915 and by 1920 was marketing all along the Eastern Seaboard under the Tydol brand. In 1931 Tidewater Oil purchased Northwestern Oil Company of Superior, WI and began expansion into the midwest. During the 1930s, control of Tidewater fell into the hands of Standard Oil of New Jersey who set up Mission Corporation as a subsidiary to operate Tidewater and Skelly. J. Paul Getty purchased control of Mission in 1937 and merged Tidewater with California's Associated Oil Company in 1938. Associated had introduced the Flying A brand in 1932 for its premium gasoline and with the creation of Tidewater Associated Oil Company, stations in the west became Associated Flying A stations and stations in the east became Tydol Flying A stations. In 1956 the Tydol and Associated designations were dropped and the brand became simply Flying A. In 1966 stations in the former Associated territory, including the secondary Seaside operation were sold to Phillips Petroleum and rebranded Phillips 66. Also in 1966 the Tidewater name was dropped in favor of Getty Oil Company, Getty having completed the purchase of Mission in 1951 and having operated Tidewater Associated and Skelly as separate subsidiaries since that time. The Getty brand first appeared in 1970, replacing Flying A at stations in the Northeast. In 1983 Getty also began replacing the Skelly brand with Getty, however before the changeover could be completed Texaco purchased Getty Oil and the Getty marketing in the Northeast was sold off to New England independent Power-Test. The former Skelly locations were rebranded Texaco or sold off and Power Test, renamed Getty Oil Marketing and based in Jericho, New York continues to operate Getty brand stations from Maine to Virginia.

ASSOCIATED OIL COMPANY - SAN FRANCISCO, CA: Founded in 1901, Associated was a West Coast integrated oil company that was marketing gasoline by 1915. In 1932 they introduced the Flying A brand as a premium grade gasoline and after the 1938 merger with Tidewater, Flying A became the company's primary brandname. Also operated stations under the Seaside brand as a secondary operation. Flying A stations in the Associated territory, as well as Seaside, were sold to Phillips in 1966.

SEASIDE OIL COMPANY - SANTA BARBARA, CA: California refiner/marketer purchased by Tidewater Associated about the end of WWII. Seaside remained a Flying A secondary brand until the Western marketing was purchased by Phillips in 1966. Phillips used the Seaside brand for a short time after the purchase but by 1973 the stations were closed or sold.

****SKELLY OIL CO. - TULSA, OK:** Skelly Oil Company was founded in Tulsa, Oklahoma in 1919 from the holdings of oil trader William Grove Skelly. The company was involved in gasoline marketing from the beginning and actively purchased chains of service stations throughout the Mid-Continent area during the early years. During the Depression Skelly defaulted on loans and control of the company fell into the hands of Standard Oil of NJ (Exxon) who formed the Mission Corporation to operate its interest in Skelly and Tidewater Oil. J. Paul Getty, in attempting to gain control of Tidewater, purchased control of Mission in 1937, only to find out he had also gained control of Skelly. Getty retained majority interest in Skelly until the 1960s when the remaining interests were purchased and Skelly was merged into Getty, along with Tidewater. The company continued to operate somewhat independently though throughout the 1970s and in 1983 Skelly began a gradual changeover to the Getty brand. Getty's purchase by Texaco halted the rebranding and most Skelly locations were ultimately rebranded Texaco by 1985.

TIDEWATER OIL CO. NEW YORK, NEW YORK

1900-1935

1915-1942

1940-1948

1948-1950

1950-1956

1956-1967

1967-1970

Getty

ASSOCIATED OIL CO., SAN FRANCISCO, CA

1910-1930s

1930s
(RED, WHITE & BLUE)

1930s

1932-1939

1938-1939

1932-1939

1939-1956

1910-1956

1950-1956

AFFILIATES
SEASIDE OIL CO., SANTA BARBARA, CALIF.

1930s

1946-1955

1955-1967

1960-1964

1960s-1970s

MOHAWK PETROLEUM
SAN FRANCISCO, CA

1930s-1970s

PATHFINDER PETROLEUM, S. PASADENA, CA

1930s-1960s

LUBRICANTS MARKETING

1910-1930s

1930s

1930s

1940-1956

1970-1983

SKELLY OIL CO., TULSA, OKLAHOMA

1923-1958

1958-1983

1975-1983

1970s SECONDARY

GETTY OIL CO., NEW YORK, NY

1970-1991

1991-PRESENT

NORTHWESTERN OIL SUPERIOR, WI

1921-1931

HOP-IN STORES, ROANOKE, VA

1970s-1980s

POWER TEST OIL WESTBURY, NY

1960s-1985

GULF

GULF OIL CORPORATION, PITTSBURGH, PA

1913-1920

1920-1932

1932-1936

1936-1954

1954-1960

1960-1963

1963-PRESENT

1956-1959
EXPERIMENTAL

SMITH OIL & REFINING CO.
ROCKFORD, IL

1920s-1950s

BRITISH AMERICAN OIL CO. TORONTO, ONTARIO

1920s-1952

1952-1967

1967-1970

ROYALITE OIL, LTD., CALGARY, ALBERTA

1953-1965

1965-1968

PETRO -CANADA
CALGARY, ALBERTA

1986-PRESENT

GULF

WILSHIRE OIL CO., LOS ANGELES, CALIF.

1920s-1950

1950s-1965

1965-TRANSITIONAL

SUNSET OIL CO., LOS ANGELES, CA

1926-1939

1939-1970s

BULK PETROLEUM CO.
CHICAGO, IL

1940s-1960

1960-1968

EXTENT OF GULF MARKETING
BEFORE EXPLOSIVE GROWTH
CIRCA 1960

CARPENTER OIL CO., LINCOLN, NE
L. L. CORYELL & SONS, LINCOLN, NE

1930s-1960s

SUPERIOR OIL CO., DES MOINES, IOWA

1930s-1960s

1960s

EARLY 1970s

PRO DEALER
NETWORK

1980s-1990s

GULF

HESS

GULF OIL CO. - PITTSBURGH, PA

The Gulf Oil Company was founded in 1901 at the Spindletop, Texas oil fields. Very early in the company's history they established a marketing presence throughout the South and East that was to dominate the company's direction for the next 80 years. Gulf is credited with having opened the first "built for that purpose, offstreet gasoline station in the world", having opened a small retail station in Pittsburgh in December 1913. Other stations predate this, but they were converted from other uses. Gulf continued to market in 28 Southern and Eastern states until the 1950s when several purchases of regional chains made Gulf a coast-to-coast marketer. The gas shortages of the 1970s forced Gulf to withdraw from many areas and the early 1980s saw Gulf in serious financial trouble. In 1984 Chevron (Standard of California) purchased Gulf Oil, converting stations in several states to Chevron and selling the southeast marketing to BP and the northeast marketing to Cumberland Farms. Cumberland continues to operate stations under the Gulf brand.

****SMITH OIL - ROCKFORD, IL:** Regional marketer purchased by Gulf in the 1950s.

****WILSHIRE OIL CO. - LOS ANGELES, CA:** Wilshire, a West Coast refiner/marketer dating from the 1910s, was purchased by Gulf in 1957. In 1965 the Wilshire brand was replaced by Gulf.

****SUPERIOR OIL CO. - DES MOINES, IA:** Superior was a Des Moines based marketer, dating from the mid-1930s. They merged with Carpenter Oil Co. in the early 1960s. In about 1968 Gulf purchased Superior and rebranded stations Gulf.

****CARPENTER OIL CO. - LINCOLN, NE:** Founded before 1900 as L.L.Coryell and Sons, Carpenter owned the Coryell brand by the early 1950s and operated stations throughout Nebraska. In the early 1960s they merged with Superior Oil of Des Moines, which was bought out by Gulf in 1968.

****BULK PETROLEUM CO. - CHICAGO, IL:** Another old Chicago private brand that sold out to a major oil company during the market expansion in the 1950s-60s. Gulf purchased the operation, branded "BULKO" in 1968. Most of the Chicago area stations were converted to Gulf, but some unsuccessful Gulf locations around the country were operated as Bulko stations in the 1970s.

****PRO:** Jobber / dealer association brand created for jobbers in the north central states when Gulf abandoned that territory. Association and brand survives today.

****BRITISH AMERICAN PETROLEUM:** Founded in 1906, British American was a fully integrated Canadian refiner and marketer. They marketed coast-to-coast and expanded greatly after the 1953 purchase of Royalite Oil. Purchased by Gulf in 1966, they became Gulf Canada in 1970 and the "B-A" brand was replaced by Gulf. Gulf Canada was sold to Petro Canada when Gulf was purchased by Chevron in 1984.

****ROAYLITE OIL LIMITED:** British-American affiliate that marketed under the Royalite brand from 1953 until 1970.

****SUNSET OIL COMPANY:** West Coast private brand purchased by Wilshire in the 1958.

****PETRO CANADA:** Purchased British American / Gulf Canada operations in the middle 1980s and continues to operate across Canada today.

****AMERADA HESS - NEW YORK, NY**

Hess began operations as a coal yard/service station in Asbury Park, New Jersey in about 1925. They evolved from coal into fuel oil during the Depression and from fuel oil into heavy residual oils for boiler fuels. Terminal facilities at Perth Amboy, New Jersey were added after WWII and the company opened it's first Hess branded retail service stations in about 1959. In 1963 they purchased independent refiner Delhi-Taylor, which owned a majority of Billups Eastern as well as Southern Oil out of Ft. Lauderdale, Florida. Southern stations were converted to Hess, and in 1965 Hess purchased the remaining portion of Billups Eastern and rebranded the stations Hess. Along the way Olixir, an Albany, New York area chain was purchased and rebranded Hess. In 1964, the first of the Christmas trucks were sold, branded Billups or Hess. Since that time there has been a promotional truck sold most every year at Christmas and these have become very desirable collectibles. In 1968 Hess Oil and Chemical merged with Amerada Petroleum, a production/exploration company with operations around the world. Hess currently markets through over 500 branded outlets in 15 states as well as suppling several strong southern private brands.

****OLIXIR - GREEN ISLAND, NY:** New York independent that sold out to Hess in the early 1960s.

****SOUTHERN OIL CO. - FT. LAUDERDALE, FL:** Florida private brand that had been purchased by Delhi-Taylor in the late 1950s. When Hess purchased Delhi-Taylor in 1963, the Southern stations were rebranded Hess.

****BILLUPS EASTERN PETROLEUM - JACKSONVILLE, FL:** When Billups was broken up in 1959, control of Billups Eastern passed to independent refiner Delhi-Taylor. In 1963 Hess purchased Delhi-Taylor and in 1965 purchased the remaining portions of Billups Eastern. Billups stations from Delaware to Florida were then rebranded Hess.

****ORLANDO ICE AND FUEL - ORLANDO, FL:** Private brand affiliated with Billups, they operated under the Spar brand in central Florida.

****MEADVILLE CORPORATION - HAVERFORD, PA:** Large chain operator of discount stations in urban Northeastern metropolitan areas. Operated under the Save Way, Safe Way and Giant brands and purchased the Merit operations based in Reading, PA. Stations continue under the Merit brand today. Hess purchased control of the Meadville Corporation in the 1960s.

****MERIT OIL COMPANY - READING, PA:** Private brand operation purchased by Meadville in the 1950s.

EXTENT OF HESS MARKETING AREA THROUGH THE YEARS

HESS

AMERADA HESS, PERTH AMBOY, NJ

1959-1980

1980-PRESENT

OLIXIR, GREEN ISLAND, NY

1930s-1962

BILLUPS EASTERN PETROLEUM JACKSONVILLE, FL

1935-1963

SOUTHERN OIL CO. FT. LAUDERDALE, FL

1950s

ORLANDO ICE AND FUEL CO., ORLANDO, FL

SPAR

1950s-1960s

MEADVILLE CORP., HAVERFORD, PA
MERIT OIL CO., READING, PA

1930s-1962

1962-1971

1971-PRESENT

1960s

1960s

1960s

HUDSON

****HUDSON OIL COMPANY - KANSAS CITY, KS:** Hudson Oil Company's storied history began in 1930 when Mary Hudson married Wayne Driver and moved to Kansas City. Mr Driver went to work with Mary's brothers at a Kansas City oil marketer operating under the Silver King name. The family was involved in two gas stations in Kansas City when, in 1933, Mr. Driver was killed in an accident with a fuel oil truck. Mary quickly sold off the two stations but within a year had purchased another station, at 25th and Broadway in Kansas City and, at age 21, was selling gasoline under the Hudson name. By 1937 the chain consisted of five stations, the original Kansas City location and stations at Leavenworth, Kansas, Norfolk and Omaha, Nebraska, and Moberly, Missouri. In the late 1930s Mary formed a partnership with her brothers and Hudson Oil was formed. The chain expanded coast to coast with hundreds of locations by the time family members went their separate ways in 1968. At that time the Hudson brothers began operating competing service stations under the Highway (Workingman's Friend) and FISCA (First In Service Company of America) brands. The 1977 purchase of a Cushing, Oklahoma refinery led to a long downward spiral for Hudson and in 1985 the Mary Hudson operation was bankrupt. More than 300 Hudson brand stations closed, leaving only a dozen or so in operation today. Hudson affiliates Highway and Fisca continue to market across the country.

HUDSON MARKETED COAST-TO-COAST

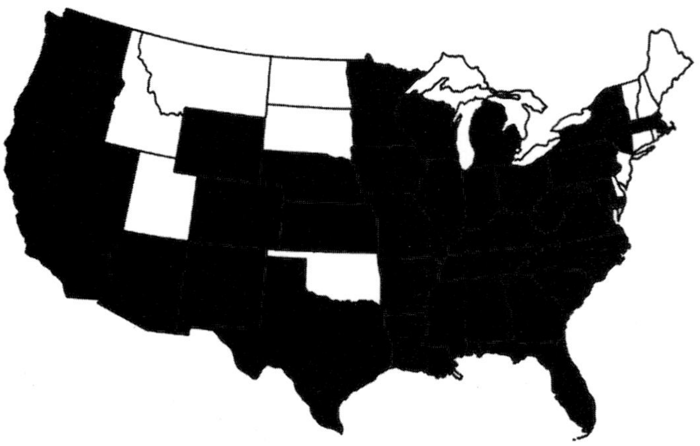

****HUDSON OF DELAWARE:** Hudson affiliate marketing under the "Hudson of Delaware" and "XCEL" brands, was purchased by Koch Industries in 1975. Koch marketing continued into the early 1980s.

****KOCH INDUSTRIES - WICHITA, KS:** Refiner and terminal operator, Koch entered retail gasoline marketing in 1975 with the purchase of the Hudson of Delaware marketing operation. Koch continued to operate the far-flung network of stations under the Hudson, Xcel, and Koch brands into the 1980s.

****HORNET OIL COMPANY - TULSA, OK:** Hornet is a Tulsa based affiliate that re-branded Hudson in the 1970s.

****HIGHWAY OIL COMPANY:** Founded in 1968 by one of the Hudson brothers, Highway Oil is perhaps the most economy minded of the various Hudson operations. Stations are small, and the company has not yet plunged into convenience store operations. They continue marketing under the Highway, Hi-Lo, and Workingman's Friend brands.

****FISCA OIL - KANSAS CITY, KS**: Also founded in 1968 by one of the Hudson brothers, Fisca is the only Hudson affiliate with stations coast-to-coast today. Fisca is an acronym for "First In Service Company of America. In recent years Fisca began adopting convenience store marketing, and many Fisca stations are operated as convenience stores today.

HUDSON

HUDSON OIL CO., KANSAS CITY, KANSAS

1933-1962

1962-1985

HUDSON OF DELAWARE

1968-1970

1970-1975

1970s

FISCA OIL CO., KANSAS CITY, KS

1968-PRESENT

HIGHWAY OIL CO. TOPEKA, KANSAS

1968-PRESENT

KOCH INDUSTRIES WICHITA, KS

1975-1980

HUDSON OIL COMPANY OF OKLAHOMA TULSA, OKLAHOMA

1955-1959

Husky

HUSKY OIL & REFINING, CODY, WYOMING & CALGARY, ALBERTA

1939-1946

1946-1954

1954-1957

1957-1960

H. EARL CLACK, INC. SPOKANE, WA

1930s-1954

FRONTIER REFINING, DENVER, CO

1937-1955

1946-1955

1955-1970

1990-PRESENT

WESTERN STATES PETROLEUM SALT LAKE CITY, UTAH

1940s

1940s-1970

1960-1978

1978-1983

1983-PRESENT

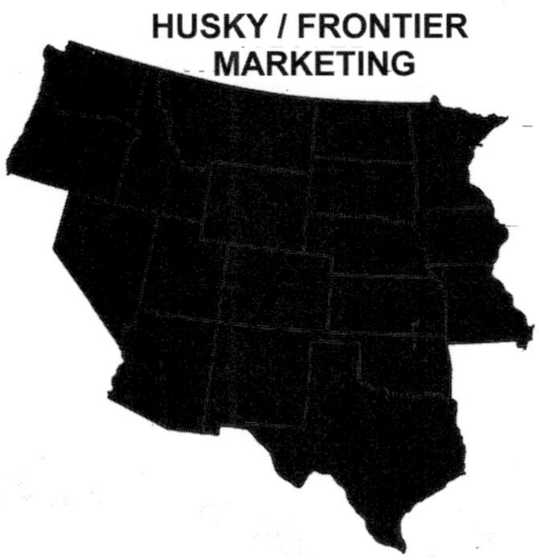

HUSKY OIL COMPANY - CODY, WY

Husky was originally a brandname of Western Oil and Fuel, Minneapolis, Minnesota. In 1938 family members associated with the founders of Western Oil and Fuel established Husky Oil Company in Cody, Wyoming. A Canadian subsidiary was created in 1946, and in 1953 Husky acquired H. Earl Clack, an independent marketer in the Rocky Mountain states. The Husky brand became Husky Hi-Power operating under the H. Earl Clack subsidiary and the brand was greatly expanded throughout the Rocky Mountain region. Western continued marketing under the Mileage name in the upper midwest, but some dealers in that area were using the Husky name as late as 1960. Husky Oil of Cody, Wyoming merged with Frontier Oil in 1968. Frontier had earlier (about 1960) bought Western States Petroleum of Salt Lake City, Utah. The Husky name replaced Frontier and Beeline about 1970 and Husky expanded into Canada. In 1984 Husky Oil sold off the U.S. marketing to concentrate on Canadian marketing. In 1990 a Denver based marketer revived the Frontier brandname and "Rarin' to Go" trademark and currently operates a chain of stations under the Frontier trademark.

****FRONTIER OIL - DENVER, CO:** Denver based refiner/marketer founded in the mid-1930s, Frontier operated a chain of stations in the Rocky Mountain region. In about 1960 they purchased Western States Petroleum and operated Beeline stations on the western slope of the Rockies. In 1968 Husky Oil purchased Frontier Oil and replaced both the Frontier and Beeline brands with Husky in about 1970. In 1990 a Denver marketer reintroduced the Frontier brandname for a chain of retail stations in Colorado.

****WESTERN STATES PETROLEUM - SALT LAKE CITY, UT:** Dating from after WWII, Western States operated the Beeline stations on the western slope of the Rockies. They merged with Frontier Oil in 1960 but the Beeline name remained in use until replaced by Husky in 1970.

DEEP ROCK OIL COMPANY, CHICAGO, IL

1920s	1915-1931	1930s	1936-1950

1950-1966	1966-1980

KERR-McGEE OIL INDUSTRIES OKLAHOMA CITY, OKLAHOMA

HARRELL BROS.

1932-1946	1946-1950	1950s CORPORATE	1960s CORPORATE

1965-1982	1965-1982	1982-1996

KNOX INDUSTRIES CORPORATION ENID, OKLAHOMA

1930s-1956	1956-1960	1960-1964

PEOPLES OIL COMPANY
NASHVILLE, TN

1930s-1946

1946-1970s

1970s

POWER OIL CO.
MACON, GA

1970s

TRACKSIDE GASOLINE STATION, INC.
GULFPORT, MISSISSIPPI

1940s-1970

1970s

TAN-KAR STATION
MONTGOMERY, AL

1950s

1970s

WALLACE SERVICE STATIONS
OF ATLANTA
ATLANTA, GA

1970s

COAST OIL CO.
SHREVEPORT, LA

1970s

MUTUAL OIL CO.
BIRMINGHAM, AL

1936-1985

TRIANGLE REFINERIES
HOUSTON, TEXAS

1930s

1940s

1947-1970s

MILEAGE MART STATIONS

1955-1960

KERR McGEE OIL INDUSTRIES - OKLAHOMA CITY, OK: Kerr McGee was founded in the Oklahoma City oilfields as Anderson-Kerr Drilling Co., in 1929. Through the early years they operated a single Oklahoma City service station, branded Phillips 66, which closed during the gas rationing days of World War II. In 1946 the corporate name was changed to Kerr-McGee, shortly after the company had entered retail gasoline marketing having purchased Oklahoma City based Harrell Bros. The Harrell Bros. "Yellow Cab" gasoline trademark was replaced by the "Kermac" brand. This was the extent of KM's gas station involvement until 1955 when they purchased Deep Rock Oil Co. Deep Rock was founded in the Cushing, Oklahoma oil fields in 1913 and went through a stormy existence, having sold out to Shaffer Oil and Refining of Chicago in the late 1920s and once again was in receivership by the late 1930s. At the time of the KM merger, the Deep Rock brand appeared on over 1,000 stations in 16 mid-continent states. In 1957 Kerr-McGee acquired Triangle Refineries, a primarily wholesale operation that was founded in 1935. Knox Industries of Texas also sold out to KM at about this time and the Knox brand survived into the 1960s. Triangle was active in purchasing Southern private brand operations which Kerr-McGee operated well into the eighties, including Peoples Oil of Nashville, 1959; Illinois based Mileage Mart in 1959; Trackside Gasoline Stations of Gulfport, Miss, 1962; Wallace Service Stations of Atlanta, 1962, along with Power Oil Co. of Macon, GA, Coast Oil Co. of Shreveport, Louisiana, and Tan-Kar Stations of Montgomery, Alabama.

****DEEP ROCK OIL CO.:** Founded in the Cushing field in 1913, Deep Rock had a series of financial problems and existed under several owners before being purchased by Kerr McGee in 1955. The Kerr McGee name was added to Deep Rock stations in about 1960 and the "brand" Kerr McGee was introduced in 1965. Deep Rock stations were gradually converted over to the KM name, nearing completion by 1985. The Deep Rock brand survives in some areas, but this is thought to be jobbers who simply haven't updated signage at smaller locations.

****KERMAC - OKLAHOMA CITY:** In 1945 when Kerr McGee purchased the former "Yellow Cab" gas stations of Harrell Bros. Oil Company the several stations were rebranded Kermac. Kerr McGee was more involved in production and refining than in marketing and the Kermac brand never appeared on more than a few stations.

****KERR McGEE:** The Kerr McGee name was added to Deep Rock stations about 1960 and the Kerr McGee brand, standing alone, first appeared on a group of Oklahoma City stations in 1965. The KM brand and logo eventually replaced Deep Rock and continued in use until 1996 when Kerr McGee abandoned branded marketing.

****KNOX INDUSTRIES - ENID, OK:** Founded in Enid in 1920, Knox operated a chain of service stations and, after 1938, auto parts and supply stores. Knox was an innovative marketer, one of the first gasoline operations to seek tie-in operations with supermarkets. Kerr McGee purchased a portion of Knox Industries in 1956, and completed the purchase in 1961. The Knox brand was eliminated in 1964 and stations were rebranded Deep Rock.

****TRIANGLE REFINERIES - HOUSTON, TX:** Founded in 1935, Triangle Refineries operated stations under the "HIOTANE" brand, as well as "TRIANGLE" and "CLOVERLEAF" and supplied various unbranded marketers. Kerr McGee purchased Triangle in 1957 and all of the private brand KM affiliates operated through Triangle.

****MILEAGE MART:** Illinois based, early gasoline c-store operation bought out in 1959 and rebranded Deep Rock.

****TRACKSIDE SERVICE STATIONS - GULFPORT, MS:** Trackside was just that--an early trackside discount operation, founded in Gulfport in the early 1930s. Kerr McGee, through Triangle, purchased Trackside in 1962 and continued to operate the chain under the Trackside name until the late 1980s.

****PEOPLES OIL CO. - NASHVILLE, TN:** Another Mid-South discount operation, Peoples dates from the 1930s and was purchased by Triangle in 1959. Many of the stations were rebranded Kerr McGee by the 1980s.

****TAN-KAR STATIONS - MONTGOMERY, AL:** Southern private brand purchased by Triangle in the 1960s.

**** POWER OIL COMPANY - MACON, GA:** Southern private brand purchased by Triangle in the 1960s.

**** COAST OIL COMPANY - SHREVEPORT, LA:** Southern private brand purchased by Triangle in the 1960s.

**** "WALCO" WALLACE SERVICE STATIONS - ATLANTA, GA:** Southern private brand purchased by Triangle in 1962.

****MUTUAL OIL CO. - BIRMINGHAM, AL:** Mutual, another private brand from the 1930s, purchased by Kerr McGee in 1985.

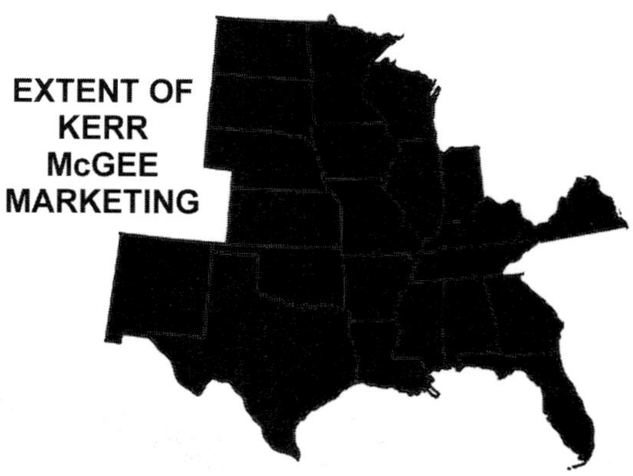

MAPCO

MAPCO PETROLEUM
MEMPHIS, TN

1980s

1990s

DELTA REFINING
MEMPHIS, TN

1950s-1980s

1990s

RED ACE PETROLEUM, NASHVILLE, TN

1933-1938

1938-1962

1962-1982

SOUTHERN INDEPENDENT OIL CO.
EVANSVILLE, IN

1940s-1970s

WESTERN GASOLINE STATIONS, INC.
NASHVILLE, TN

1960s-1970s

****MAPCO PETROLEUM - TULSA, OK AND MEMPHIS, TN:** Created in the 1970s in a consolidation of several regional gasoline marketers, Mapco markets under the Mapco and Delta brands in the South Central states today.

****DELTA OIL COMPANY - MEMPHIS, TN:** Memphis marketer purchased by Mapco.

****RED ACE PETROLEUM - NASHVILLE, TN:** Founded in 1933 as one of many southern discount chains, Red Ace came under the control of a holding company that also purchased Delta and Western. The operation was sold to Mapco about 1981, and continues to operate under the Mapco brand today.

****WESTERN GASOLINE STATIONS - NASHVILLE, TN:** One of many southern discount chains, Red Ace came under the control of a holding company that also purchased Delta and Western. The operation was sold to Mapco in about 1981, and continues to operate under the Mapco brand today.

****SOUTHERN INDEPENDENT OIL COMPANY - EVANSVILLE, IN:** Regional marketer founded in 1934 and purchased by Mapco in the 1970s.

Marathon best in the long run

MARATHON OIL CORPORATION - FINDLAY, OH: The Ohio Oil Company was founded in 1887 when several northern Ohio oil production firms joined together to market their crude oil, primarily to Standard Oil. So much so that in 1889 Standard purchased Ohio Oil and operated the company as a production subsidiary until the breakup of Standard in 1911. After the breakup, Ohio Oil continued as before with oil exploration and production, satisfied to sell their product as crude oil and leave the refining and marketing to others. Ohio Oil first ventured into marketing with the June 1924 purchase of Lincoln Oil Refining Co. of Robinson, Illinois. Lincoln Oil operated a refinery and about 20 retail gasoline stations in the Robinson-Terre Haute area at the time of the purchase by Ohio Oil. With an assured supply of crude oil for the small refinery, the Linco brand began to expand. By 1930 Ohio Oil had expanded the Linco brand throughout most of the current Marathon territory—Ohio, Indiana, Illinois, Michigan and Kentucky. In the meantime, Ohio Oil had been involved in numerous successful oil production ventures and found they needed even more outlets for their products. In 1930 Ohio Oil purchased Transcontinental Oil, a refiner/marketer that had marketed gasoline under the trademark "MARATHON" since about 1920. They had acquired the brand from a Pittsburgh operation, Riverside Oil Co., that we believe to be a forerunner of Republic Oil Company, which will be mentioned later. Transcontinental sold gasoline under the Marathon trademark across the Midwest and South, from North Carolina to New Mexico. Transcontinental can best be remembered for a significant first when in 1929 they opened several Marathon stations in Dallas, Texas in conjunction with Southland Ice Company's "Tote'm" stores (later 7-Eleven), creating the first gasoline - convenience store tie-in. The Marathon brand proved so popular that by WWII the name had replaced Linco at stations in the original five-state territory. During the War, the widespread Marathon territory couldn't be properly supplied and marketing operations outside the Great Lakes area were sold, primarily to Tydol. Concentrating their marketing in a smaller territory allowed Marathon to expand within the territory and the late 1950s brought the first of many acquisitions, Tower Oil Company of Cincinnati, Ohio, that would expand the company's marketing presence. Numerous other purchases are detailed in the listings below, with a group of marketing companies brought together by the 1980s under a subsidiary, Emro Marketing. Ohio Oil, renamed Marathon Oil in 1962, was purchased by U.S. Steel (now USX) in 1981 and continues to market through dealer and jobber outlets under the Marathon name in the North Central states and through Emro Marketing under several brandnames, primarily Speedway and Starvin Marvin, throughout the South and East.

****LINCOLN OIL REFINING CO. - ROBINSON, IL:** Purchased by Ohio Oil in 1924. Linco served as the company's primary brand until about 1939.

****RED FOX OIL COMPANY - FT. WAYNE, IN:** Regional marketer purchased by Linco in 1928 and rebranded Linco at that time.

****TRANSCONTINENTAL OIL CO. - TULSA, OK:** Purchased by Ohio Oil in 1930, the Transcontinental/Marathon marketing was kept separate from Linco marketing until 1936 and Marathon had completely replaced Linco by WWII.

****TOWER OIL CO. - CINCINNATI, OH:** Tower was a Cincinnati based marketer that sold out to, and rebranded Marathon in 1958.

****AURORA GASOLINE CO. - DETROIT, MI:** Aurora Gasoline Company was a Detroit refiner marketer that introduced the Speedway brand in the early 1930s. The company's premium product, Speedway 79 Stratofuel, was introduced before WWII and the primary brandname used from the end of WWII until 1962 was "Speedway 79". In 1959 Marathon purchased the Aurora Gasoline Company and the Speedway 79 logo was redesigned. Gas pump globes were eliminated at this time. The Speedway 79 brand was replaced by Marathon in 1962, however, in 1975, the Speedway brandname was reintroduced at several company owned retail "gas only" stations. Marathon began buying up regional private brands and consolidating them under Emro Marketing with the Speedway brandname. As of this writing, the Speedway name appears on petroleum products retailed at hundreds of Emro "Starvin' Marvin" c-stores.

****OLD DUTCH REFINING CO. - MUSKEGON, MI:** Founded in Muskegon in about 1930, the Old Dutch brand appeared on a small number of stations, primarily jobber operated, in the Muskegon area. As best as can be determined, Aurora Gasoline Co. purchased Old Dutch Refining in the late 1950s, although jobbers kept the Old Dutch brand alive well into the 1970s.

****REPUBLIC OIL CO. - PITTSBURGH, PA:** Dating from early 1923 and a creation of Michael Benedum and Joseph Trees, Republic was the successor to Riverside Oil Co. (originator of the Marathon trademark). Republic supplied branded and unbranded jobbers throughout the Southeast. Republic marketing was concentrated in small areas and no attempt was made at complete statewide coverage in the states where they marketed. By the 1950s the corporate name was Plymouth Oil Co. and Marathon purchased the operation in 1962. Concentrating on the extensive unbranded market, the Republic brandname was eventually phased out, although some jobbers continue to use the brandname today.

****HI-WAY:** Republic secondary brand.

****BONDED - SPRINGFIELD, OH:** The Springfield, Ohio based affiliate of Gaseteria-Bonded of Indianapolis. Dating from 1932, Bonded of Ohio remained independent after Gaseteria-Bonded sold out to Esso's Oklahoma division in 1957. Marathon purchased Bonded in January 1975 and although many locations were converted to Speedway/Starvin'-Marvin, the Bonded brand has remained in use at many locations.

****CONSOLIDATED - OSHKOSH, WI:** A Wisconsin discount operation from the postwar era that operated about 30 stations throughout the state. Purchased by Marathon in July 1972 and rebranded Speedway in about 1982.

****GASTOWN OF DELAWARE:** Cleveland area marketer purchased Marathon in August 1971.

****CHEKER OIL CO. - CHICAGO, IL:** Cheker was a suburban Chicago collection of private brands, the Cheker name originally replacing "Chief" and "Road Chief" in about 1968. No globes are known from these brands. Cheker purchased the Sun-Glo operation in Indiana in about 1970 and converted those stations to the Cheker brand. In 1976 Cheker, then 50% owned by Exxon, purchased the Imperial Refineries operation. Imperial, founded in St. Louis in 1915, was in its earliest days an integrated oil company with production, refining and marketing. After WWII, Imperial resumed marketing as a discount chain operation, eventually operating stations in nearly 20 states. Cheker operated Imperial Refineries as a separate brand until the September 1983 merger with Marathon/Emro and Emro replaced most of the Imperial brand with Ecol after the 1984 purchase of this Jackson, MS independent.

****IMPERIAL REFINERIES - CLAYTON, MO:** Founded in 1915 as a refiner/marketer, Imperial operated discount stations throughout the Midwest and South until the early 1980s.

****JOHNSON OIL CO. - MT. PLEASANT, MI:** Founded in 1926 in Lakeview, Michigan and currently Imperial Oil Co. of Mt. Pleasant, Michigan—was apparently an Imperial jobber during the time Imperial Oil had refineries and a jobber marketing network. They continue to market in Michigan and Indiana under the Imperial name. This company is not related to Marathon marketing, and is listed here only because of the ties to Imperial.

****PORT OIL COMPANY - CHARLESTON, SC:** Regional independent that marketed in North and South Carolina. Purchased by Marathon / Emro in August 1984.

****VALUE - WEBSTER SERVICE STATIONS - GREENVILLE, SC:** Regional independent that marketed in North and South Carolina. Purchased by Marathon / Emro in January 1984.

****ECOL - JACKSON, MS**: Regional independent that marketed throughout the Southeast. Purchased by Marathon / Emro in May 1984.

****GLOBE OIL COMPANY - CHAMBLEE, GA:** Regional independent that marketed in Georgia and North and South Carolina. Purchased by Marathon / Emro in December 1985. Globe Oil was the originator of the Starvin' Marvin convenience store operation, and the name was adopted for use by all of the Emro operation stores.

****ROCK ISLAND REFINING - INDIANAPOLIS, IN:** Founded in Duncan, Oklahoma in the 1920s, Rock Island was a mid-continent area marketer until WWII. Having shifted to wartime refining and virtually abandoning retail gasoline, the company was purchased by the Koch family in 1946 and moved to Indianapolis, where it became a major source of unbranded gasoline for independent marketers. Over the years the company acquired several Indiana and Michigan marketers including Colonial/Progressive of Ft.Wayne, Tulsa of Detroit, United of Indianapolis, Naph-Sol of Muskegon and Golden Imperial (listed below) of Indianapolis. By the mid 1980s "United" had become the primary brand, replacing the others where convenient. In 1989 Marathon's Emro Marketing purchased Rock Island and many United locations have converted to the Speedway brand.

****GOLDEN IMPERIAL - INDIANAPOLIS, IN:** Small regional independent purchased by Rock Island in the 1970s.

****COLONIAL/PROGRESSIVE - FT. WAYNE, IN:** Independent marketer founded in the 1930s that operated Colonial stations in Indiana and Progressive stations in Michigan. Sold out to Rock Island in the 1960s.

****TULSA OIL CO. - DETROIT, MI:** Tulsa is an old Detroit independent brand that sold out to Crystal Refining in the early 1970s and was in turn purchased by Rock Island in about 1978. The United brandname replaced Tulsa in 1985.

****UNITED - INDIANAPOLIS, IN:** United was an Indianapolis based discount marketer with stations throughout Indiana and a few in Michigan. In 1985 United was purchased by their primary supplier, Rock Island, and became the brandname used to consolidate all of Rock Island's marketing. Most of the United stations were rebranded Speedway in 1990 but some locations remain United.

****NAPH-SOL REFINING CO. - MUSKEGON, MI:** Naph-Sol entered the gasoline market with the introduction of Zephyr gasolines in 1932. Marketing continued through jobbers and independent dealers in Michigan and Wisconsin until 1989 when Naph-Sol was bought out by Rock Island Refining. Several Michigan jobbers continue to use the Zephyr brand.

****OSCEOLA REFINING CO. - REED CITY, MI:** Independent refiner/marketer purchased by United Refining in 1973. Naph-Sol purchased the operation after the 1970s petroleum shortages and continued to operate the brand, in limited markets, until the mid-1980s.

****J.A. HOGSHIRE OIL COMPANY, INDIANAPOLIS, IN:** Founded in 1926 and an original member of the Independent Oil Men of America, Hogshire originally operated under the Criteria brand. In 1937 the Wake Up brand was introduced. Marathon, through Emro, purchased Wake Up in the early 1990s and continues to brand stations with the Wake Up name today.

****KWIK SAK - NASHVILLE, TN:** Regional convenience store chain purchased by Emro in the mid 1990s.

****MARTIN OIL SERVICE - ALSIP, IL:** Martin Oil Service is the division that expanded coast-to-coast beginning in the late 1950s. Sold out to Marathon's Emro division in 1993. Emro continues to operate stations under the Martin brand in Illinois and Indiana.

Marathon — best in the long run

LINCOLN OIL REFINING, ROBINSON, IL

1918-1924 | 1924-1936 | 1936-1939

TRANSCONTINENTAL OIL, TULSA, OK

1915-1926 | 1915-1926 | 1926-1936 | 1936-1939

THE OHIO OIL CO., FINDLEY, OH

1936-1946 | 1946-1962 | 1962-PRESENT

TOWER OIL CO. CINCINNATI, OH

1930s-1958

RED FOX OIL CO. FT. WAYNE, IN

1920-1928

OLD DUTCH REFINING MUSKEEGON, MI

1932-1971

AURA GASOLINE COMPANY, DETROIT, MI

1930s

1930s

1935-1950

1946-1952

1952-1959

1959-1962

1975-1986

1986-PRESENT

1980s EXPERIMENTAL

REPUBLIC OIL CO. PITTSBURG, PA

1929-1946

1946-1970s

REPUBLIC SECONDARY BRAND

1960s-1980s

CONSOLIDATED GASOLINE STATION OSHKOSH, WI

1940-1965

1965-1982

GASTOWN OF DELAWARE

1960s-1980s

1980s

BONDED OIL
SPRINGFIELD, OH

1950-1970

1970-PRESENT

CHEKER OIL CO.
CHICAGO, IL

1968-1984

1984-PRESENT

IMPERIAL REFINERIES CLAYTON, MISSOURI

1915-1940

1946-1976

1960s
EXPERIMENTAL

1976-1980

1980s

JOHNSON OIL CO.
MT. PLEASANT, MI

1946-1980s

1980s

PORT OIL CO.
CHARLESTON, SC

1970s-PRESENT

CHEKER AFFILIATE

1970s-1980s

WEBSTER SERVICE STATIONS
GREENVILLE, SC

1980s

ECOL
JACKSON,MISSISSIPPI

1980s

KWIK SAK
NASHVILLE,TN

1970s-PRESENT

GLOBE OIL CO. CHAMBLEE, GA

1970s

1980s

1980s

1990s

ROCK ISLAND REFINING INDIANAPOLIS, IN

1930s-1946

1946-1960s

COLONIAL-PROGRESSIVE FORT WAYNE, IN

1950s-1970s

GOLDEN IMPERIAL OIL INDIANAPOLIS, IN

1960s-1970s

UNITED OIL CO. INDIANAPOLIS, IN

1980s-PRESENT

1960s

1980s-PRESENT

TULSA OIL CO. DETROIT, MI

1930s

1950s-1970s

NAPH-SOL REFINING. MUSKEGON, MICH.

1931-1946 | 1946-1957 | 1957-1960 | 1960-1989 | 1989

OSCEOLA REFINING CO., REED CITY, MI

1946-1960 | 1960-1989

J. A. HOGSHIRE, INC.
INDIANAPOLIS, IN

1926-1937 | 1937-1962 | 1962-1980s | 1980s-PRESENT

MARTIN OIL SERVICE, ALSIP, ILLINOIS

1956-1961 | 1956-1961 | 1961-1962 | 1962-PRESENT

MARTIN

MARTIN OIL CO.,
CARBONDALE, ILLINOIS

SMARTIN UP WITH MARTIN
MARTIN
OIL PRODUCTS

1930s-1977

J. D. STREET & CO.,
MARYLAND HEIGHTS, MO

ZEPHYR — **1939-1946**

USE ZEPHYR OIL AND GASOLINE — **1946-1954**

MOTOR OIL ZEPHYR GASOLINE — **1954-1960**

ZEPHYR — **1960-PRESENT**

MARTIN OIL SERVICE, ALSIP, ILLINOIS

SAVE ON MARTIN ETHYL Wonderful SERVICE — **1956-1961**

MARTIN OIL SERVICE, INC. — **1956-1961**

MARTIN — **1961-1962**

MARTIN — **1962-PRESENT**

MARTIN OIL COMPANY - CARBONDALE, IL
MARTIN OIL SERVICE - BLUE ISLAND, IL

Martin, an Illinois based trackside discounter dating from the early 1930s was actually a collection of several family owned operations within the Martin chain. In the earliest days dual headquarters--Chicago and Carbondale, Illinois--were listed on products and later the Chicago division was moved to Alsip and still later Blue Island, Illinois. In about 1956 the chain separated into two distinct operations, the Chicago (Alsip) division becoming "Martin Oil Service" while the Carbondale division remained Martin Oil Company. Martin Oil Service expanded coast-to-coast, operating over 200 gas stations by the 1970s while Martin Oil Company concentrated marketing through a small number of stations in the south central states. Cut off from supplies during the 1973-74 gas shortage, Martin Oil Service retreated to their home base, currently operating less than 50 stations in Illinois and Indiana. Martin Oil Company sold out to J.D. Street of Maryland Heights, MO in 1977 and the stations were rebranded with J.D. Street's Zephyr brand. In the 1980s J.D. Street became a branded jobber for several majors and the Zephyr brand remains in use on only a small number of stations today.

****MARTIN OIL STATIONS - CARBONDALE, IL:** Operated stations in the South Central states from the 1930s until purchased in 1977 by J.D. Street.

****J.D. STREET - MARYLAND HEIGHTS, MO:** Founded in 1884 as a lubricants distributor, J.D. Street entered the retail petroleum market with the introduction of the Zephyr gasoline brand in 1939. J.D. Street purchased Martin Oil Co. in 1977 and converted their stations to the Zephyr brand. Currently serving as a jobber for several major brands as well as operating a few remaining Zephyr brand stations.

****MARTIN OIL SERVICE - ALSIP, IL:** Martin Oil Service is the division that expanded coast-to-coast beginning in the late 1950s. Sold out to Marathon's Emro division in 1993, and Emro continues to operate stations under the Martin brand in Illinois and Indiana.

MIDWEST OIL COMPANY
MINNEAPOLIS, MN - FARGO, ND - SIOUX FALLS, SD

1919-1927 | 1927-1930 | 1930-1946

1946-1960 | 1955-1960 | 1960-1977

AVIO - MIDWEST SECONDARY BRAND

1930s | 1940s-1960

CORPORATE

1930s

H. K. STAHL, INC., MINNEAPOLIS, MN

1930s | 1940-1958

****MIDWEST OIL COMPANY - SIOUX FALLS, SD:** Founded in Minneapolis in 1919, Midwest marketed under the "Ace High", "Ace", and "Avio" brands in the upper Midwestern states. Branches were established in Sioux Falls, SD and Fargo, ND in the 1920s, and in 1977 Midwest sold off their Minneapolis and Fargo operations to concentrate on the Sioux Falls branch. The company continues there today as a branded Sinclair marketer.

MIDWEST MARKETING AREA

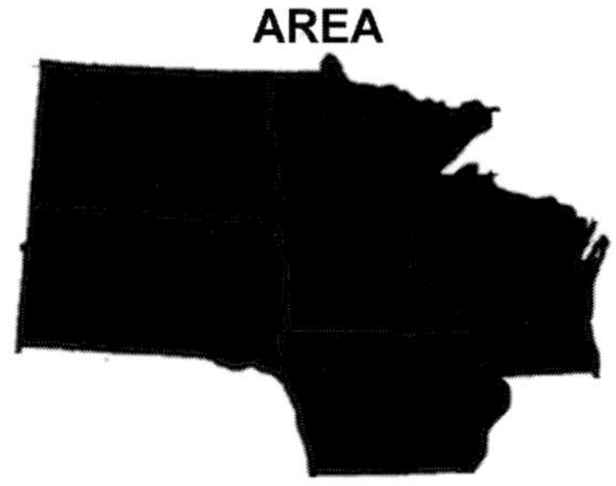

****H. K. STAHL, INC - MINNEAPOLIS, MN:** Independent marketer whose gasoline operations were purchased by Midwest in 1958.

Mobilgas

****MOBIL OIL CORPORATION - FAIRFAX, VA**
****STANDARD OIL COMPANY OF NEW YORK/SOCONY - NEW YORK, NY:** Founded in 1882 as the administrative division of the Standard Oil Trust, Socony eventually became involved in refining and later marketing in New York and New England. With the breakup of Standard in 1911, Socony was assigned its refining and marketing properties but left without crude oil production. Exploration began, but most of the crude oil production came through acquisition of Magnolia Petroleum in 1918 and General Petroleum in 1926. Socony expanded it's marketing through Magnolia, General, and White Eagle, reaching from coast-to-coast by 1930. In 1931 Socony merged with another Standard affiliate, Vacuum Oil, the new corporate entity to be named Socony-Vacuum Oil Company. Socony-Vacuum mated Socony's shield and flying horse with Vacuum's "MOBILGAS" trademark and expanded the Mobilgas brand to all of the affiliated companies by the end of WWII. In 1955 the company was renamed Socony-Mobil and in 1966 Mobil Oil Corporation. See SOCONY-VACUUM section below for more information.
****MAGNOLIA PETROLEUM - DALLAS, TX**: A Texas oil producer/refiner/marketer dating from 1894, Magnolia sold 45% interest to Socony in 1918. Socony completed the purchase in 1925. The Magnolia brandname was replaced by "MOBILGAS" in about 1934, after the Socony-Vacuum merger. Magnolia remained a separate, named subsidiary, primarily involved in pipeline transportation until 1960.
****GENERAL PETROLEUM CORPORATION - VERNON, CA:** A pioneer California oil producer and refiner, General was founded in 1910. In 1926 Socony purchased General Petroleum and maintained the General brands until the end of WWII, gradually replacing them with Mobilgas. As with Magnolia, General remained a separate, named subsidiary until 1960.
****WHITE EAGLE OIL AND REFINING - KANSAS CITY, MO:** Mid-Continent producer/refiner/marketer with gas stations in 11 Midwestern states. Chief White Eagle was a leader of the Ponca Indians in Oklahoma. White Eagle was first used as a corporate name in 1916 when L.L. Marcell founded the White Eagle Petroleum Company. That same year the new company built a refinery at Augusta, Kansas, having a capacity of 2,000 barrels per day.

In 1919 the White Eagle Petroleum Company was reorganized and the name changed to the White Eagle Oil and Refining Company. This reorganization was in reality a merger of six independent oil companies and within a few years the new company enjoyed second position in the distribution of petroleum products in the Middle West. A second refinery was built at Casper, Wyoming, and one was purchased at Fort Worth, Texas.

The poised White Eagle was adopted as an emblem because of its dignity and general significance, particularly in their territory and because it was in keeping with the company name. As a service station identification, this emblem—the big iron eagle in front of the station and the illuminated glass eagles on the pumps—was very effective.

In January 1930, the White Eagle Oil and Refining Company was purchased by the Standard Oil Company of New York (Socony) at which time the name was changed to the White Eagle Oil Corporation. When Socony and Vacuum merged in 1931, White Eagle Oil Corporation became a subsidiary company. In 1935 the White Eagle Oil Corporation was dissolved and the operation became the White Eagle division of Socony-Vacuum Oil Company, Inc.
****VACUUM OIL COMPANY - ROCHESTER, NY:** Founded as a lubricants manufacturer in 1866, Vacuum had developed an extensive line of specialty lubricants when Standard Oil purchased the company in 1879. During the Standard era, Vacuum's lubricants, under the Gargoyle and later Gargoyle Mobiloil brands, were sold by the various Standard marketing firms. With the breakup of Standard in 1911, Vacuum was left to market its oils through independent channels worldwide. They became involved with gasoline marketing, under the "MOBILGAS" brand, in the late 1920s and expanded the gasoline venture with the purchase of Lubrite Refining Co. of St. Louis, Missouri in 1929. The following year Vacuum purchased Wadhams Oil Corporation of Milwaukee and the Vacuum "MOBILGAS" name was added to Wadhams extensive product line. Also in 1930 Vacuum purchased White Star Refining of Detroit and added over 1,500 outlets for Mobilgas. In 1931 Socony and Vacuum merged to form Socony Vacuum Oil Company and after a two year reimaging program, the Mobilgas brandname became the company's primary brand, although several of the regional marketers retained their own identities until after WWII.
****LUBRITE REFINING COMPANY - ST. LOUIS, MO:** Regional marketer purchased by Vacuum Oil in 1929.
****WADHAMS OIL CORPORATION - MILWAUKEE, WI**: Founded in the 1870s as a lubricants manufacturer, Wadhams was a pioneer in gasoline marketing, operating Wadhams stations throughout Wisconsin prior to 1920. Wadhams was briefly affiliated with the Independent Oil Men of America in 1925 and expanded with the 1929 purchase of Milwaukee competitor Bartles-Maguire Oil Company. Noted for its Chinese pagoda style stations, Wadhams had developed a first-class gasoline marketing organization when the company was purchased by Vacuum in 1930. Vacuum's Mobilgas trademark was added to the Wadhams brands and finally replaced the Wadhams name in about 1939.
****BARTLES-McGUIRE OIL CO. - MILWAUKEE, WI:** Large Wisconsin marketer that merged with Wadhams in 1929, shortly before Vacuum purchased Wadhams.
****WHITE STAR REFINING CO. - DETROIT, MI:** Also added to Vacuum's Mobilgas marketing was Detroit's White Star Refining. White Star operated over 1,500 stations in Ohio, Indiana and Michigan when purchased by Vacuum in 1930. Mobilgas replaced the White Star brand almost immediately, but former White Star outlets used a special Mobilgas shield sign with a small blue circle and white star at the bottom until after WWII.
****INDEPENDENT OIL CO. OF PENNSYLVANIA - ALTOONA, PA:** Purchased by Socony-Vacuum in 1934 and rebranded Mobilgas.

METRO - JAMESTOWN/OLEAN, NY: Small New York marketer purchased by Vacuum in 1928. The Mobilgas brand was added at Metro stations (note the red/white/green "Mobilgas" globe in the Vacuum section) and Metro became a low-grade product sold at Mobil stations in the various marketing territories until after WWII.

GILMORE OIL CO. - LOS ANGELES, CA: Founded in 1900, Gilmore entered gasoline marketing in 1923. The company established an elaborate network of dealers in California, Oregon and Washington, totaling more than 3,500 outlets at its peak in the late 1930s. In 1940 Socony-Vacuum purchased control of Gilmore and assigned Gilmore marketing to the General Petroleum subsidiary. Transition from Gilmore to Mobilgas began in 1942 and was completed in 1945, just after WWII.

MOBILGAS/SOCONY-VACUUM: As is noted in the introduction, Socony and Vacuum merged in 1931. By 1934 it was decided to join the Socony shield and flying horse with Vacuum's Mobilgas to create a single brandname for all the affiliates. Several, including Magnolia and White Eagle, readily converted to the new Mobilgas identity. Others, including General and Wadhams, combined the Mobilgas brand with their own brandnames, at least for the first few years. Conversion was nearly complete when WWII interrupted, and was completed as soon as possible after the war.

**MOBIL - Although the Mobil name replaced Mobilgas as the primary station identification on a gradual basis between 1955 and 1958, gasoline brands remained Mobilgas and Mobilgas Special until 1962 when they were replaced with Mobil Regular and Mobil Premium. Mobil again reimaged in 1966, replacing the elongated Mobil shield with the modern "red O" logo.

RESPECTIVE AREAS OF MOBIL AFFILIATES

STANDARD OIL COMPANY OF NEW YORK NEW YORK, NY

1911-1928

1920-1933

1931-1940

MAGNOLIA PETROLEUM, DALLAS, TX

--1920s-1931--

1931-1940s

Mobilgas

GENERAL PETROLEUM CORP., LOS ANGELES, CALIFORNIA

1918-1931

WHITE EAGLE OIL & REFINING

1920s

1920s

1926-1930

1930-1931

VACUUM OIL COMPANY

VACUUM OIL COMPANY

1926-1933

LUBRITE OIL CO., ST. LOUIS, MO

PRE-1929

1929-1933

WHITE STAR REFINING, DETROIT, MI

PRE 1930

1930-1933

1933-1946

WADHAMS OIL CO., MILWAUKEE, WI

1911-1919

1919-1940

1926-1933

1933-1946

BARTLES BONDED GASOLINE
MILWAUKEE, WI

PRE 1929

METRO GAS STATIONS, INC.
JAMESTOWN/OLEAN, NY

PRE 1932

INDEPENDENT OIL CO
ALTOONA, PA

PRE-1934

GILMORE OIL CORP.
LOS ANGELES, CA

1923-1946

INDEPENDENT PETROLEUM CO.
SEATTLE, WA AND PORTLAND, OR

1940s

Mobilgas

1931-1948

1948-1955

1955-1966

MOBIL OIL CORPORATION, McLEAN, VA

1966-PRESENT

MOBILOIL LUBRICANTS

1901-1931

1931-1946

1946-1955

1950s

1950s

1955-1966

MOBIL SECONDARY BRANDS

1970s-1980s

1970s-1980s

PENNZOIL CORP. & SOUTH PENN OIL CO., OIL CITY, PA.

GASOLINE

1921-1936

1936-1940s

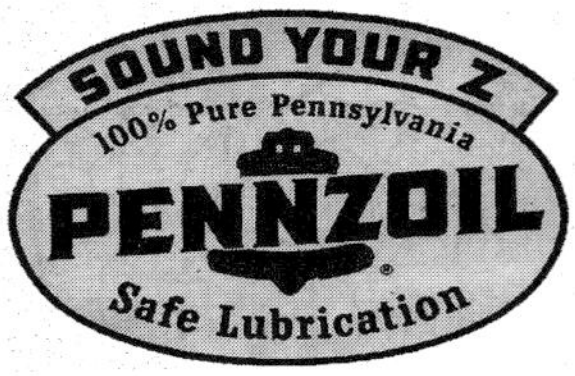

1930s

1930s

LUBRICANTS

1918-1921

1921-1928

1928-1940

1930s-1950s

WOLVERINE-EMPIRE REFINING OIL CITY, PA

GASOLINE

1928-1940

1930s-1970s

LUBRICANTS

PRIOR TO 1926

1932-1949

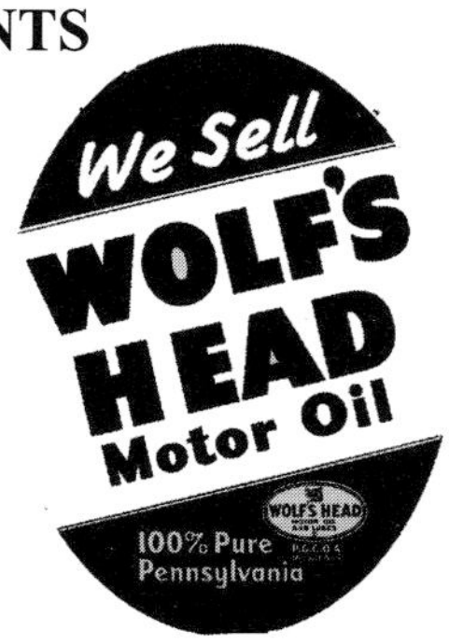

1949-1973

1973-PRESENT

1938-1950

1950-1957

1957-1962

1962-1980

1980s-PRESENT

1935-1950

1950s-1970s

1970s

ELK REFINING CO.
CHARLESTON, WV

1920s-1930s

1930s-1963

FLEETWING CORPORATION
CLEVELAND, OH

1968-PRESENT

PENNSYLVANIA REFINING CO.
BUTLER, PA

1920-1929

1929-1973

1973-PRESENT

PENNZOIL COMPANY - HOUSTON TX/OIL CITY, PA: Pennzoil traces it's origins to the founding of South Penn Oil Company, a production unit of the Standard Oil Trust. When South Penn became independent with the breakup of Standard in 1911, they remained primarily a production operation. In 1925, they purchased control of The Pennzoil Company, an Oil City refiner and lubricants marketer that had in 1921 opened the first of a small chain of branded Pennzoil stations. The gasoline marketing was rebranded "Pennzip" in 1936 and remained such until 1958 when the Pennzoil brand was reintroduced for gasoline. South Penn purchased the remaining interest in Pennzoil in 1955 and the company was renamed Pennzoil Company in 1963. Numerous purchases detailed below greatly expanded the scope of the company and today the company sells gasoline through 600+ stations in eight states.

In 1984 Pennzoil attempted to purchase Getty Oil, only to have Texaco interfere and prevent the merger from taking place. Pennzoil sued, and the awarded damages brought Texaco to the brink of extinction, breaking Texaco into two smaller units. Pennzoil suddenly enriched, has continued to expand and modernize its facilities, and renewed the focus on marketing of Pennzoil branded lubricants in the United States and around the world.

****WOLVERINE-EMPIRE REFINING CO. - OIL CITY, PA:** Founded in 1879 as Empire Oil Works, Wolverine-Empire was primarily a lubricants manufacturer. They entered the gasoline market under the Empire brand about 1928 and rebranded Wolf's Head, after their famous motor oils, in the late 1930s. In 1963 Pennzoil purchased Wolf's Head and continued to operate the Wolf's Head stations as a secondary brand into the 1980s. Wolf's Head motor oils and lubricants continue to be a favorite in many markets today.

****ELK REFINING CO. - CHARLESTON, WV:** Founded in 1913 as an affiliate of United Refining Company of Warren, PA, Elk was the refiner/marketer that operated the Keystone stations in West Virginia. South Penn purchased control of Elk in 1952 and the remaining portion in 1963. The Pennzoil brand replaced the Keystone name in about 1965.

****FLEET WING CORPORATION - CLEVELAND, OH:** Pennzoil purchased the Fleet Wing operation from Sohio in 1970 and continues to operate a small number of Fleet-Wing branded stations today. See Sohio section for more Fleet-Wing information.

****PENNSYLVANIA REFINING COMPANY - BUTLER, PA:** Founded in Butler in 1878; Pennsylvania Refining, later Pennreco, marketed gasoline and lubricants under the Penn-Drake brand after the 1929 purchase of American Gasoline Company and their "Penn-Drake" brand. The company withdrew from gasoline marketing at the start of WWII and later, in 1973, was purchased by Pennzoil Company. They continue to market a line of specialty oils today.

****PHILLIPS PETROLEUM - BARTLESVILLE, OK:** Phillips Petroleum was founded in 1917 in Bartlesville, Oklahoma from various production and refining companies owned by Frank Phillips and others. The company entered gasoline marketing in November 1927 when they opened a small station in Wichita, Kansas. Other stations were quickly added and other marketing operations were purchased and by 1930 the company operated over 6,000 stations in 12 states. Details on some of the companies purchased are listed in individual company histories, below. Phillips was primarily a mid-continent marketer until 1953 when they opened their first stations in Florida, setting off a marketing chain-reaction that saw them contracting with dissatisfied jobbers from other marketers and rebranding stations to Phillips 66 as far north as Virginia. Pleased with the market expansion, Phillip's "drive to Maine" expansion began, eventually pushing the brand all the way through the north Atlantic region and on to Maine. Never before had an oil company attempted such a systematic market expansion. With the purchase of Tidewater's west coast Flying A stations in 1966, Phillips found themselves operating stations in 49 states. A single unit in Alaska was opened in 1967, making Phillips the second marketer to brand stations in all 50 states at the same time (Texaco was first and Shell has been in all 50 states but not at the same time). The 1970s gas shortages forced Phillips to abandon many areas of the country, but currently the Phillips 66 brand appears on nearly 8,000 stations in 29 Mid-continent and Southeastern states.

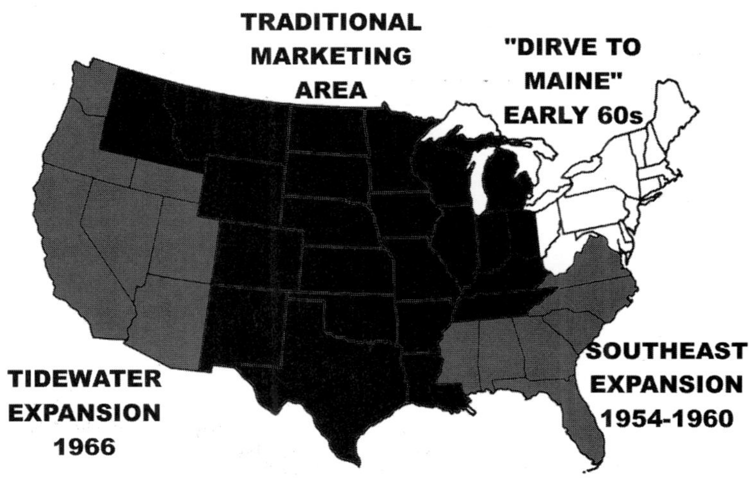

****BENZO GAS MOTOR FUEL CO. - KANSAS CITY, MO:** Kansas City marketer purchased by Phillips in about 1928.

****THE INDEPENDENT OIL AND GAS CO. - TULSA, OK:** Successors to Waite Phillips Oil Co., merged into Phillips Petroleum in 1930.

****MANHATTAN GASOLINE CO. - ST. PAUL, MN:** Manhattan originated the motor oil trademark "TROP-ARTIC" in 1912 and marketed Trop-Artic oils through Manhattan stations in the North Central states until Phillips purchased Manhattan in 1934. Stations were at that time rebranded Phillips 66, but Phillips adopted the Trop-Artic motor oil brand and continues to use it today.

****THE QUAKER OIL CO. - ST. LOUIS, MO:** Quaker Oil sold the Paraland gasoline brand from the early 1930s until after WWII. The brand shifted to Paraland Oil Co. of Omaha, Nebraska by the 1950s and it is not known if Phillips owned the company at that time. By the early 1960s it became known that Paraland was Phillips' "secret" secondary brand. In later years unsuccessful Phillips stations were converted to Paraland. The brand was last used in about 1972.

****WASATCH OIL CO. - SALT LAKE CITY, UT:** Wasatch was a Salt Lake City based collection of refining and marketing companies that branded stations under the Wasatch trademark in Utah, and various "Chief" brands representative of the state name in Idaho, Montana, Oregon and Washington. Phillips purchased Wasatch in 1947 and stations were rebranded Phillips 66 at that time. Golden Eagle stations in California, originally associated with Wasatch, were sold off as independents, and continued to operate as early self-service discounters until purchased by Ultramar.

****HOME OIL COMPANY - BUTTE, MT:** Founded in 1936, they marketed under the Silver Gas brandname until purchased by Wasatch in 1947, about the time of the Phillips - Wasatch merger.

****SEASIDE OIL:** See Getty section.

PHILLIPS PETROLEUM CO., BARTLESVILLE, OK

1927-1928

1928-1959

1929-1934

1930-1959

1959-1980

1966-1980

1980-PRESENT

INDEPENDENT OIL TULSA, OK

1926-1929

1929-1932

BENZO GAS MOTOR FUEL CO. KANSAS CITY, KS

1920s

MANHATTAN OIL CO, ST. PAUL, MN

1910s-1934

EXPORT

1980s

PACIFIC PETROLEUM, CALGARY, ALBERTA

1960-1981

QUAKER OIL COMPANY

PREWAR

1946-1953

1953-1972

WASATCH REFINING & INLAND EMPIRE REFINING
SALT LAKE CITY, UTAH

USED IN WA, OR, MT, ID

USED IN UTAH

USED IN RESPECTIVE STATES AS LISTED

1930s

1940s

PANAMA REFINING

1940s-1960

HOME OIL COMPANY
BUTTE, MT

1936-1947

SEASIDE OIL CORP.
SANTA BARBARA, CA

1966-1973

PURE

THE PURE OIL COMPANY

1917-1920 | 1920-1930 | 1927-1930 | 1930-1954

WOFFORD OIL CO., ATLANTA, GA

1910s-1926 | 1926-1940

W. H. BARBER & CO., MINNEAPOLIS, MN

1915-1930s | 1926-1946 | 1946-1965 | 1941-1965

CENTRAL WEST OIL CO. SOUTH BEND, IN

1950s | 1950s-1970

WISCONSIN INDEPENDENT OIL CO. MILWAUKEE, WI

1940s | 1950s

PURE

1938-1954

1954-1970

1957-1964

1964-1970

1961-1970

1954-1970 and 1992-PRESENT

1957-1970

HICKOCK OIL CO., TOLEDO,OH

1930s

1917-1952

POCAHONTAS OIL CO. CLEVELAND, OH

1930-1936

1936-1940

1936-1940

PURE

THE PURE OIL CO. USA - CHICAGO, IL: Pure Oil was founded in 1891 as the Producers Oil Co. in Oil City, Pennsylvania. The name was changed to The Pure Oil Company in 1895. Gasoline marketing began in 1914, and in 1917 ownership of the company passed to the Dawes family who operated public utility companies in Ohio and elsewhere. By the early 1920s the utilities had been sold off and the company was firmly in the gasoline business. In the 1920s Pure grew through a series of mergers with other regional marketers including Wofford Oil Co. of Atlanta, Georgia; Sherrill Oil Co of Pensacola, Florida; Seaboard Oil Company of Jacksonville, Florida; and the American Oil Co. of Hattiesburg, Mississippi. Also in the late 1920s the company introduced their famous "cottage" service station design that was to dominate Pure marketing forever and influence the competition until after WWII. In 1965 Pure Oil merged with Union Oil of California and began a five-year program to replace the Pure brand with Union 76. The last of the Pure signs came down in 1970, thus passing into history one of the most distinctive images in gasoline marketing.

****WOFFORD OIL CO. - ATLANTA, GA:** In 1925 Pure Oil merged with Wofford Oil Company of Georgia and Wofford Oil Co. of Alabama. Wofford had marketed WOCO-PEP, a benzol blend, for a number of years and had used the seal type logo that Pure Oil adopted in 1930. Purol branded stations throughout the South continued to market "WOCO-PEP" until WWII.

****HICKOCK OIL CO. - TOLEDO, OH:** Founded in 1917, Hickock Oil Co. marketed gasoline under the HI-SPEED brandname. Pure purchased 51% of Hickock Oil Co. in 1928. Hickock operated the "HI SPEED" stations throughout Ohio and Michigan, and were the originators of the successful Truck Stop concept that was greatly expanded upon by Pure after the purchase of Hickock was completed in 1951 and their stations and truck stops were rebranded Pure.

****POCAHONTAS OIL COMPANY - CLEVELAND, OH:** Local independent purchased by Hickock in 1935.

****W.H. BARBER AND CO. - CHICAGO, IL:** Founded in Minneapolis in the 1910s, they operated the Fyre Drop stations and later the Meteor stations from about 1925 on. They introduced the SAVE MORE brand in 1940. W.H. Barber also owned the Central West/Super Par operation at the time they were bought out by Pure in 1945. W.H. Barber was spun off as an independent after the Union/Pure merger in 1965 and markets lubricants today.

****SAVE MORE:** W.H. Barber discount brand introduced in 1940 for company operated locations.

****CENTRAL WEST OIL CO. - SOUTH BEND, IN:** A subsidary of W.H. Barber. Operated the Super Par and Wisco 99 stations.

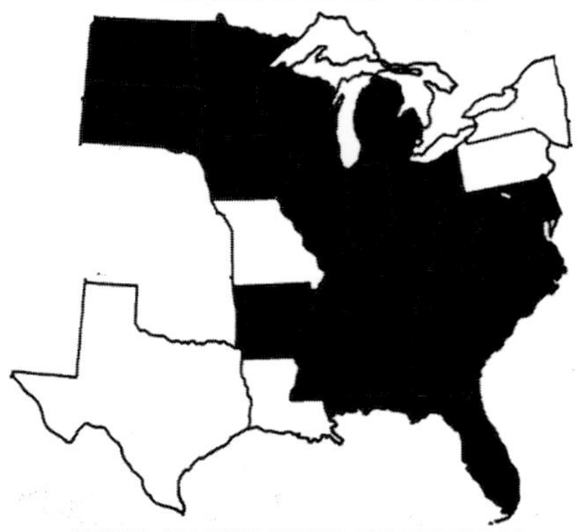

GASOLINE MARKETING

1923-1946

1946-c1960

1960s

1970s

1980s

LUBRICANTS

1930s

1940s

1950s

1960s

1960s-1970s

1980s-PRESENT

Quaker State

Diamond Shamrock

QUAKER STATE OIL REFINING CO. - OIL CITY, PA: In 1931 Sterling Oil Company of St. Mary's, West Virginia, Quaker State Oil Refining Company of Oil City, Pennsylvania and 16 other independent oil producers, refiners, lubricant compounders and marketers joined together to form Quaker State Oil Refining Corporation. It was decided that Sterling would remain the gasoline brand as Sterling was the only company involved with any significant gasoline marketing. The primary motor oil brand would be Quaker State, originally introduced in 1914 and distributed nationwide through lubricant jobbers, often jobbers of competing gasoline brands. The merger was the first time the famous motor oil brandname had been tied to any particular gasoline brand and it was thought that marketing gasoline under a different name would lessen the problems associated with gasoline jobbers selling a "competitors" motor oil. Sterling would remain the gasoline brandname until after 1960 when stations were rebranded Quaker State. Quaker State gasoline is currently sold through a small number of stations in Pennsylvania and Ohio.

****STERLING OIL COMPANY - ST. MARY'S, WV:** Pennsylvania producer/refiner/marketer that originated the Pennsylvania Grade Crude Oil Association. Sterling was involved in gasoline marketing by the early 1920s and remained the brandname for gasoline marketed by Quaker State when that company entered the gasoline market after merging with Sterling and others in 1931.

EXTENT OF STERLING / QUAKER STATE BRANDED GASOLINE MARKETING

RESPECTIVE SHAMROCK AND ULTRAMAR MARKETING AREAS TODAY

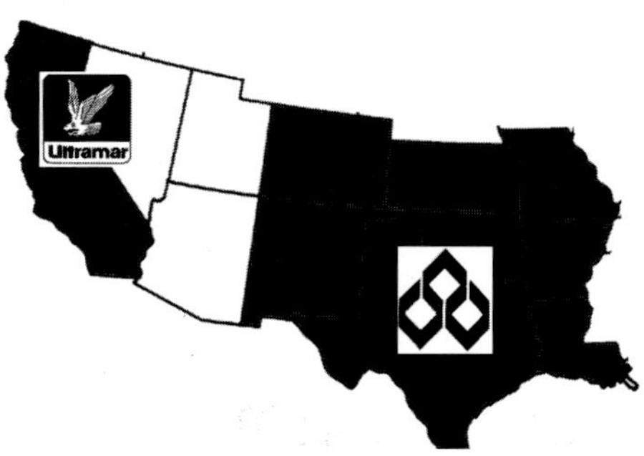

DIAMOND SHAMROCK, INC. - SAN ANTONIO, TX: Diamond Shamrock, founded in 1927 as Shamrock Oil and Gas Corp. in Amarillo, Texas, has long been a regional marketer in the South Central and Southwestern U.S. In 1967 Shamrock Oil and Gas merged with Diamond Alkali Company, a chemical manufacturer, to form Diamond Shamrock, and operations were moved to San Antonio. The company has gradually expanded it's market in recent years and now brands over 2,000 locations. Shamrock has historically marketed through jobbers, several including Sigmor (purchased by Shamrock in 1983) and Cliff Brice, have become large marketing organizations themselves. In 1996 Ultramar PLC, the British corporation owning Ultramar Canada and Ultramar, Inc. (the former Caminol / Beacon operation) purchased Diamond Shamrock and merged the operations to form Ultramar Shamrock. The Diamond Shamrock brand survives in the South Central and Southwestern states, and the Ultramar brand survives in Canada and California. As this publication went to press, Ultramar Diamond Shamrock, as the new concern is known, announced plans to purchase all of the domestic marketing properties of Total Petroleum. The Ultramar, Shamrock, and Total brandnames will be retained in their respective areas.

****CLIFF BRICE - PUEBLO, CO:** Cliff Brice is a Pueblo, Colorado based Shamrock regional jobber operating dual branded Cliff Brice/Shamrock stations.

**** FILL 'EM FAST - HOUSTON, TX:** The Fill 'em Fast brand was introduced in 1968 by Autotronics, Inc., a Houston firm specializing in gasoline installations at existing convenience stores. In the early 1970s Sigmor Corporation, a large regional Shamrock jobber, purchased Autotronics and expanded the Fill 'em Fast brand with a series of company operated pumpers throughout the Southeast. With the buyout of Sigmor by Shamrock in the late 1980s, the far-flung Fill 'em Fast operation was abandoned, and individual stations were sold locally.

****BEACON: THE CAMINOL COMPANY-HANFORD, CA:** Founded in 1931, The Caminol Company operated stations under the Beacon brand. The company became Beacon Oil Company after WWII and introduced an early version of the gasoline convenience store to the West Coast in about 1958. They were purchased by Ultramar in 1981 and stations continue to operate under both the Beacon and Ultramar brandnames today.

****ULTRAMAR:** Founded in England in 1935 as an exploration company, they purchased Panama Refining's Golden Eagle stations in California in 1960. In 1969 they purchased Canadian "Herts & Bets Oil Co." and rebranded to Golden Eagle. They purchased Beacon Oil in 1981. In 1996 Ultramar merged with Diamond Shamrock to form Ultramar Shamrock, and in 1997 the new entity announced plans to acquire all domestic marketing properties of Total Petroleum. They plan to continue operations under the Shamrock brand in the South Central states, Total in the Midwest and North Central states, and the Beacon and Ultramar brandnames on the West Coast.

Diamond Shamrock

SHAMROCK OIL & GAS CORPORATION AMARILLO, TX
DIAMOND SHAMROCK, SAN ANTONIO, TX

1930s   1946-1962   1962-1980s   1980-1993   1993-PRESENT

CLIFF BRICE PETROLEUM
PUEBLO, CO

1950s   1960s-PRESENT

AUTOTRONICS, INC.
HOUSTON, TX

1968-1985

THE CAMINOL CO. - BEACON OIL CO.
HANFORD, CALIFORNIA

1930s   1940s   1950s-1970s   1970s   1980s

PANAMA REFINING

1940-1960   1960-1981

ULTRAMAR, PLC

1981-PRESENT

Shell

SHELL OIL CO.
NEW YORK, NY & HOUSTON, TX

1912-1922

1922-1928

1928-1946

1946-1960

CANADIAN OIL CO., LTD.
TORONTO, ONTARIO

1917-1925

1925-1936

1936-1952

1952-1964

1964
TRANSITIONAL

NORTH STAR OIL LTD.
WINNEPEG, MANITOBA

1930s-1956

1956-1961

Shell

1946-1960 | 1960-1964 | 1960-1977 | 1977-PRESENT

SHELL OIL COMPANY-HOUSTON, TX: Shell Oil Company, a unit of the British/Dutch "Royal Dutch Shell" entered U.S. markets with the purchase of a terminal in Washington state in 1912. In 1917 Shell purchased St. Louis based Roxana Petroleum and by 1929 was marketing in all 48 states. Since that time Shell has proved to be the most innovative marketers, often leading the way in new gasoline marketing concepts. Several Shell "firsts" include the 1947 introduction of internally lit plastic signage, the 1957 introduction of landscaped "rancher" style stations, and one of the first franchised Auto Care programs. Shell continues to market through company direct and jobber operated locations in 41 states and around the world today.

NORTH STAR PETROLEUM - WINNIPEG, MANITOBA: Independent Canadian refiner - marketer founded in 1920. They were purchased by Shell in 1961.

CANADIAN OIL COMPANIES, LTD. - TORONTO, ONTARIO: In 1936 National Refining sold off their Canadian marketing and from it Canadian Oil Companies, Ltd. was formed. They continued to market in Canada under the White Rose brandname until purchased by Shell and converted to the Shell brand in 1964.

SIGNAL

SIGNAL OIL CO. - SIGNAL OIL & GAS CO.
LOS ANGELES, CA

1932-1962

1948-1970
CORPORATE

1963
EXPERIMENTAL

1963-1967

HANCOCK REFINING COMPANY LONG BEACH, CA

1926-1934

1934-1940

1940-1959

1959-1970

1959-1965

1965-1970

BANKLINE OIL CORP.

1930s-1959

1959-1965

1950s-1970

REGAL SERVICE STATIONS, SAN JOSE, CA

1950s-1960s

SIGNAL OIL AND GAS - LOS ANGELES, CA: In 1932 Signal Oil and Gas of Los Angeles, formerly a producer/refiner entered the retail marketplace under the Signal name. Involved with Standard of California from its earliest days, Signal sold its retail marketing to Standard Chevron in 1947. In 1958 Signal merged with Hancock Oil of Long Beach, CA and bought the Houston based Watson Oil and rebranded the stations "Hancock". The Signal brand was operated by Chevron until about 1965 when the entire marketing network was sold to Humble Oil and stations were rebranded Enco. Hancock, Norwalk and other Signal brands were sold off beginning in 1970. Stations in the West were sold off individually, while the Southeastern marketing was sold to Charter Oil/Dixie Vim.

****HANCOCK REFINING CO. - LONG BEACH, CA:** Founded in 1925, Hancock was a California based refiner/marketer with stations throughout the state. In 1958 Signal Oil and Gas merged with Hancock Refining and stations remained under the Hancock brand in California and the brand was expanded to Texas when Signal purchased the Watson operation and rebranded the stations Hancock. Hancock brand discontinued when Signal marketing was sold off in 1970.

****WATSON OIL CO. - HOUSTON, TX:** Houston based discounter purchased by Signal in about 1960 and rebranded Hancock.

****REGAL PETROLEUM - SAN JOSE, CA:** California independent purchased by Signal in about 1960.

****BILLUPS BROS. - GREENWOOD, MS:** Founded in Greenwood, Mississippi in 1926, Billups survived as a family owned operation until 1959. In 1959 the company was split into two operating units, Billups Eastern and Billups Western. Billups Western was sold out to Signal Oil and operated by them until 1970 when Signal got out of retail gasoline marketing and sold the operation to Charter Oil. A majority of Billups Eastern was sold to Independent refiner Delhi-Taylor, which was bought out by Hess Oil and Chemical in 1963. The remaining portion of Billups Eastern was bought by Hess in 1965 and Billups sta-

SIGNAL

BILLUPS WESTERN PETROLEUM
GREENWOOD, MS

1935-1988

1971-1988

ROSE OIL CO., VICKSBURG, MS

1940s-1970s

SUPERTEST OIL CO., TAMPA, FL

1940s-1960s

1959-1965

1960s-1988

1988-PRESENT

SOUTHLAND OIL CO.
SAVANNAH, GA

1940s-1987

WATSON OIL CO.
HOUSTON, TX

1950s-1960s

OLD COLONY OIL
CHELSEA, MA.

1970s

SOUTHERN OIL STORES, INC.
JACKSONVILLE, FL

1928-1946

1946-1960

1960-1988

CHARTER OIL
JACKSONVILLE, FL

1980s

tions were rebranded Hess. See Hess section for more information.

****ROSE OIL CO. - VICKSBURG, MS:** Rose Oil Co. was apparently operated by a member of the Billups family.

****SUPERTEST OIL CO. - TAMPA, FL:** Postwar Florida based discounter bought out by Signal in about 1960 and sold off to Charter in 1970.

****SOUTHLAND OIL CO. - SAVANNAH, GA:** In 1970 when Signal Oil sold off its retail marketing, "SOC"--Southland Oil Co. of Savannah, GA--was sold off to its original owners, who had remained in management since Signal's purchase of SOC about 1960. They operated as an independent until about 1986 when they became a branded Texaco operation.

****SOUTHERN OIL STORES/DIXIE VIM - JACKSONVILLE, FL:** In 1970 Signal Oil again left the retail gasoline marketplace. Charter Oil Co. of Jacksonville, FL had the year before purchased the Southern Oil Stores-Dixie Vim chain and when Billups, Supertest and the former Watson/Hancock stations were placed on the market, they added them to the Dixie Vim operation. Southern Oil Stores was an independent marketer, one of the original trackside discount chains with approximately 50 stations. Founded about 1929 it is listed here only to aid in understanding the change in ownership of Billups and Supertest.

****CHARTER OIL - JACKSONVILLE, FL:** Formed in 1969 from Dixie Vim, they became a regional marketer in the 1970s before dissolving in the mid-1980s. They also operated the Old Colony stations in New England in the 1970s.

SINCLAIR

1916-1920

1920-1926

1920-1926

1926-EXPERIMENTAL

SINCLAIR EXPERIMENTAL IMAGES

1953-1956

1957-1960

SINCLAIR EXPLORATION

1920-1970

PIERCE OIL, S. A.
(MEXICO)

1930-1938

CUBA

1926-1959

SHERWOOD BROTHERS, BALTIMORE, MD

1919-1939

1940-1964

1940-1953

1946-1964

SINCLAIR

1926-1959 | 1959-1970 | 1970-1973 | 1973-PRESENT

PRODUCERS & REFINERS
INDEPENDENCE, KS

1920s-1934

PIERCE PETROLEUM
ST. LOUIS, MO

1920-1927 | 1927-1930

RICHFIELD OIL COMPANY OF NEW YORK
NEW YORK, NY

1929-1937 | 1937-1940 | 1940-1964

STOLL REFINING CO.
LOUISVILLE, KY

1910s-1930s | 1930s-1952

SIMPSON OIL CO.
ST. LOUIS, MO

1946-1964

SINCLAIR

PASCO MARKETING
PHOENIX, AZ

1960s

LITTLE AMERICA REFINING
LITTLE AMERICA, WY

1930s-1950s

SINCLAIR OIL CORPORATION - NEW YORK, NY: Founded in 1916 in New York from the various holdings of founder Harry Sinclair, including the refining and marketing formerly under the Cudahy name, Sinclair was a huge operation right from the start. With oil production, pipelines, railcars, refineries, terminals, bulk plants and service stations, Sinclair was one of the few oil companies to start as a completely integrated operation. Sinclair grew larger during the prosperous 1920s in spite of Harry Sinclair's involvement in the "Teapot Dome" oil scandal of President Warren Harding's administration. With several important purchases during the early 1930s, such as Pierce, Parco, Rio Grande and Richfield of New York, Sinclair was marketing from coast-to-coast. With the 1937 reorganization of Rio Grande into Richfield of California, Sinclair's direct operations on the West Coast were discontinued, but Sinclair continued to market from the Rockies east. From 1932 until 1943 the corporate name was Consolidated Oil Company, changing back to Sinclair Oil Corporation in 1943. After WWII Sinclair continued to expand direct operated and jobber marketing, building thousands of service stations along the new interstate highway system. Marketing surpassed production and the company could no longer supply all the crude oil needed for its own branded marketing. In 1969 Sinclair merged with oil-rich and newly created Atlantic-Richfield, selling off its East Coast marketing to British Petroleum. In 1973 Atlantic-Richfield was court-ordered to sell most of Sinclair, and the operation was spun off as an independent oil refiner and marketer that today operates more than 2,500 stations in 20 states.

****PIERCE PETROLEUM CORPORATION - ST. LOUIS, MO:** Primarily a production and refining affiliate of the Standard Oil Trust. Operated as Waters-Pierce Oil Company prior to 1920. Had extensive gasoline marketing in Mexico. Entered gasoline marketing in the south central states in the early 1920s. Purchased by and rebranded Sinclair in 1930.

****PRODUCERS AND REFINERS OIL COMPANY - INDEPENDENCE, KS:** Rocky Mountain refiner/marketer partially owned by Prairie Oil and Gas. When Sinclair purchased Prairie in the early 1930s, they also acquired part ownership of PARCO. They purchased the remaining portions in 1934 and the operation was rebranded Sinclair.

****LITTLE AMERICA REFINING COMPANY - LITTLE AMERICA, WY:** A Wyoming refiner/marketer that had ties to Fletcher Oil Co. The Larco brand was purchased by Sinclair in 1955. Larco marketed under the Covey and Mobile brandnames in the 1950s.

****SIMPSON OIL CO. - ST. LOUIS, MO:** Missouri marketer purchased by Sinclair in 1961. The Simpson "SUPER A" brand survived as a Sinclair secondary brand until 1965.

****STOLL REFINING CO. - LOUISVILLE, KY:** Louisville refiner/marketer founded in 1896 and involved in Kerosene and lubricants in the early years of this century. Entered gasoline marketing in the 1920s and continued supplying stations under the Stoll "Golden Tip" brand until after the 1948 purchase of Stoll refining by Sinclair. The "Golden Tip" stations were rebranded Sinclair in about 1952.

****RICHFIELD OIL COMPANY OF NEW YORK - NEW YORK, NY:** Richfield, a West Coast producer/refiner/marketer was founded in California in 1901. By 1915 Richfield was marketing on the west coast. The company entered east coast markets in the mid-1920s, and created a subsidary, Richfield Oil Co. of New York with the 1929 purchase of Walburn Petroleum Company. During the 1930s, as detailed below, Richfield was reorganized and the Eastern marketing was sold to Sinclair.

****SHERWOOD BROS. - BALTIMORE, MD:** Large Richfield/Sinclair jobber operation which had marketed under the Betholine trademark prior to branding Richfield and continued to dual brand "Betholine/Richfield" and "Betholine/Sinclair" until the late 1950s.

****PASCO MARKETING - PHOENIX, AZ:** Southwestern independent that operated a portion of the Sinclair marketing in the interim between control by Atlantic Richfield and the recreation of the current Sinclair Oil Corporation.

SOHIO

STANDARD OIL CO. OF OHIO - CLEVELAND, OH: Sohio is, of course, Standard Oil of Ohio, which traces its roots to the original formation of Standard Oil by John D. Rockefeller in Cleveland, Ohio in 1870. Although Standard of New Jersey was to become the "main office" as such after the formation of the Standard Oil Trust in 1882, Standard of Ohio was always considered to be "the original". After the break up of Standard Oil in 1911, Standard of Ohio was assigned only marketing in Ohio and a small refinery. Being primarily a marketing company, Sohio was a pioneer in service stations, operating a chain of company owned stations statewide by 1920. The brandname "SOHIO" was introduced in 1928 to identify all the company's marketing. In their first marketing venture outside Ohio, Sohio purchased Spears and Riddle of Wheeling, West Virginia, a large regional marketer selling gasoline under the Fleet-Wing name. Fleet-Wing Corporation was formed as a subsidiary and the brand was offered to jobbers in the states surrounding Ohio. Sohio opened the first company owned locations outside Ohio in Newport, Kentucky in 1956 under the Boron name. Boron had first been used as a brandname for Sohio's super-premium grade of gasoline in 1954. Little was done with the Boron brand until stations were opened in Western Pennsylvania in 1962. In 1969 British Petroleum purchased a controlling interest in Sohio and BP's U.S. marketing subsidiary was transferred to Sohio. British Petroleum had entered Canadian markets in the 1950s and purchased Supertest Oil in 1967. About that time Sohio also purchased Petroleum Marketing, Inc. of McLean, Virginia. Boron and BP became Sohio's primary non-Ohio brandnames. Fleet-Wing was sold to Pennzoil in 1968. In 1990 BP, having purchased the remaining portions of Sohio, began rebranding stations under the Boron, Wm. Penn, Gas-n-Go, and Gulf (Southeast, purchased in 1985) to BP. Sohio stations were rebranded BP beginning in 1991 and now the entire company markets under the BP name.

****CALDWELL & TAYLOR - CINCINNATI, OH:** Cincinnati based marketer purchased by Sohio in 1928.

****SPEARS & RIDDLE - WHEELING, WEST VIRGINIA:** A large regional wholesale operation founded in the 1870s, selling gasoline under the Fleet-Wing brandname. Sohio purchased Spears & Riddle in 1928 and created the Fleet-Wing Corporation as a marketing subsidiary to brand jobbers in Ohio and surrounding states. Pennzoil purchased Fleet-Wing from Sohio in about 1970 and continues to operate about a dozen stations under the Fleet-Wing name.

****THE REFINERS OIL COMPANY - DAYTON, OH:** Refiners was a Dayton, Ohio based marketer that operated around 300 retail stations in Ohio and Indiana. When the General Motors Chemical Company developed the antiknock additive Tetraethyl Lead in early 1923, Dayton based Refiners was chosen to test market the new product. The test was so successful that in September of that year Standard Oil of Indiana signed on to be an exclusive distributor of the new additive, "Ethyl", through its marketing territory. Standard Oil of Ohio purchased Refiners Oil company in 1930 and rebranded the stations Sohio.

****CANFIELD OIL COMPANY - CLEVELAND, OHIO:** Founded in Cleveland in 1886, Canfield was primarily a lubricants manufacturer, developing, among others, the "Wm. Penn" motor oils. Canfield became involved in gasoline marketing in the late 1910s, operating stations in and around Cleveland. When Standard Oil purchased Canfield in 1945, there were over 200 branded stations at that time. The Sohio brandname gradually replaced the Canfield name, although Canfield and Wm. Penn motor oils continued in production. In the 1970s, after British Petroleum purchased a controlling interest in Standard Oil, many dealer operated BP stations in the East were rebranded "Wm. Penn", continuing as such until the BP name was selected to replace all Sohio affiliate brandnames in 1990.

THE STANDARD OIL COMPANY (OHIO)
CLEVELAND, OH

RED CROWN GASOLINE

1912-1928 (GREEN)
1928-1930 (BLUE)

SOHIO

1928-1936

SOHIO

1936-1940

SOHIO STANDARD OIL

1940-1952

SOHIO

1940-1962

SOHIO

1962-1985

SOHIO

1985-1992

SOHIO

TRUCKSTOPS OF AMERICA

1950s-PRESENT

BORON OIL COMPANY
CLEVELAND, OH

1956-1962

1962-1992

CANFIELD OIL CO., CANFIELD, OHIO

1880s-1930s

1930-1965

1930s-1940s

1965-1980s

SPEERS & RIDDLE
WHEELING, WV

1920s-1930

1930-1940

FLEET-WING CORPORATION
CLEVELAND, OH

1940-1960s

1960s

GAS & GO
SECONDARY BRAND

1980s

PETROLEUM MARKETING CORP., McLEAN, VA

1950-1972

1972-1980

SOHIO

REFINERS OIL CO., DAYTON, OHIO

1916-1923

1923-1927

1927-1930

CALDWELL AND TAYLOR CINCINNATI, OH

1920s

BP AMERICA, CLEVELAND, OH

1969-1988

1969-1988

1988-PRESENT

BP CANADA, MONTREAL, QUEBEC

1959-1980s

SUPERTEST OIL CO. LONDON, ONTARIO

1926-1955

1955-1974

GIBBS OIL CO., BOSTON, MA

1967-1970s

1970s-1980s

1960s

SPUR DISTRIBUTING COMPANY, NASHVILLE, TN

1928-1940 | 1938-1940 | 1939-1962

RIVER STATES OIL COMPANY, EVANSTON, IL

1955-1957 | 1957-1961

DIRECT OIL COMPANY, NASHVILLE, TN

1946-1962

INGRAM OIL COMPANY, NASHVILLE, TN

1956-1961

WEBB OIL COMPANY, DULUTH, MN

1923-1930s | 1930s-1956 | 1956-1960s | 1960s

EXTENT OF SPUR BRAND

WOOD RIVER OIL AND REFINING, WOOD RIVER, IL

1939-1950 | 1950-1970s

MURPHY OIL CORPORATION, EL DORADO, ARKANSAS

1961-1976 (RED FIELD)	1976-1980 (BLUE FIELD)	1980-1992	1992-PRESENT

1976-1980

HEMPSTEAD OIL STORAGE COMPANY, MOBILE, AL

1980-1992

VIGOR OIL COMPANY AN ONTARIO MARKETER

MURPHY OIL CORPORATION

CORPORATE 1961-PRESENT

1950s

PRE-1961

1964-1996

****MURPHY OIL CO. - EL DORADO, AR:** Murphy traces it's origins to the banking, timber and oil interests of founder C.H. Murphy, when brought together in 1907. Oil production began in earnest in the 1920s when company shifted almost entirely to oil exploration after WWII. In 1957 Murphy entered the retail market with the purchase of River States Oil Company of Evanston, IL. Murphy then introduced the Royal brandname and was expanding successfully when the company purchased Spur Distributing in 1960. By 1962 all the former Royal stations had rebranded Spur and the company added several other marketers to the collection, replacing that operation's brand with Spur. The company brands over 500 Spur stations in 14 states today.

****RIVER STATES OIL CO. - EVANSTON, IL:** River States, marketing in Wisconsin, Minnesota, Illinois and Michigan, was a combination of several North Central marketing firms brought together about 1955. In 1957 River States was purchased as the first marketing division of Murphy Oil and the "ROYAL" brand was introduced, replacing the earlier "RIVER STATES" brand. In 1962 the Spur brandname replaced Royal at several hundred stations in the North Central states.

****WEBB CUT PRICE GAS AND OIL - DULUTH, MN:** Founded about 1925, Webb operated stations in Minnesota, Wisconsin and Michigan until about 1970 when they either merged with or rebranded Spur.

****SPUR DISTRIBUTING, NASHVILLE, TN:** Founded by John Mason Houghland in Nashville in 1928 as the original trackside gasoline chain, Spur had 348 stations in 21 states when they were purchased by Murphy Oil in September 1960. Murphy, who had only recently entered the retail marketing field, marketed under the various regional names (Royal, Spur, Ingram) until 1962 when the Spur brandname was chosen as the company's only brand and the others were replaced with the new oval and spur logo. The company supplies about 500 branded stations in 14 states today.

****WOOD RIVER OIL COMPANY - WOOD RIVER, IL:** Formed in 1939 as a refiner/marketer. The Ingram family was involved with Wood River from its founding and eventually Wood River operations in the South were spun off and rebranded Ingram.

****INGRAM OIL - NASHVILLE, TN:** The Ingram brand was introduced at Wood River stations in the South in 1956. Murphy Oil purchased the Ingram marketing, 201 stations, in 1961 and rebranded the stations Spur.

****HEMPSTEAD OIL STORAGE CO. - MOBILE, AL:** Purchased by Murphy Oil in 1962 and rebranded Spur.

SUNOCO

SUN OIL CO., PHILADELPHIA, PA

1920-1946

1939-1959

1959-1962

1962-PRESENT

MID CONTINENT PETROLEUM, TULSA, OK

1920s-1930s

1930s-1946

TRANSITIONAL
1940s

1933-1957

1957-1990s

SUNRAY OIL CO., TULSA, OKLAHOMA

1930s

1930s

1930s-1955

1955-1957

1957-1962

PREMIER OIL CO., LONGVIEW, TEXAS

PREMIER

1955-1960s

PREMIER

1960s-1968

PREMIER/FREEWAY

1968-1973

PREMIER/FREEWAY

1973-1990s

Freeway

1960s

BARNSDALL REFINING CO. BARNSDALL, OK

BARNSDALL BE B SQUARE PRODUCTS

1921-1940

BARECO B Dependable Products

1940-1957

RESPECTIVE MARKETING AREAS AT TIME OF 1968 MERGER

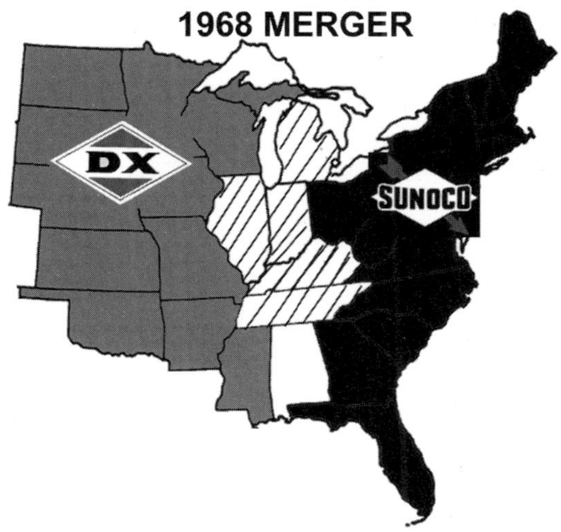

MONARCH OIL CO., COUNCIL BLUFFS, IA

SEARLE PROCESS MonaMotor BEST BECAUSE PUREST

1918-1930

MASCOT STATIONS - SECONDARY BRAND

Mascot

1970s-1980s

MID STATE PETROLEUM LEXINGTON, NC

TRAVELERS GASOLINE

1940-1958

TRAVELERS

1958-1964

TRAVELERS

1964-1970s

TRAVELERS

1968-PRESENT

TRAVELERS

1992-PRESENT

SUNOCO

KENDALL REFINING, BRADFORD, PA

GASOLINE

LUBRICANTS

1920s-1976

1955-1976

1928-1940

1940-1970s

WAVERLY OIL WORKS

1920s-1930s

WITLO/AMALIE OIL COMPANY

1920s

1930s

1946-1970s

1970s-PRESENT

****SUN OIL COMPANY:** Sun Oil, a refiner-marketer founded by the Pew family in Philadelphia in 1890, began retail gasoline marketing in 1920. Primarily a Northeastern marketer, Sunoco had stations from Maine to the Chicago area south to Virginia by the early 1930s, as well as marketing in Florida. For many years Sunoco marketed only a premium grade of gasoline, "Blue Sunoco", but for a brief time after WWII added a super premium "Sunoco Dynafuel". In 1968 Sun merged with Sunray DX and for the next twenty years marketed under the Sunoco name in the Northeast and East while retaining the DX name in the Mid-continent area. In about 1988 Sun began re-imaging in the Midwest, replacing the DX brand with Sunoco at high volume locations and selling off or closing the others. Some pockets of DX branding remain, but this is primarily from jobbers who just haven't removed old signage.

****MASCOT and TRAVELERS:** Sunoco secondary brands, Travelers was originally a brand used by Taylor Oil of Winston-Salem, NC. It shifted to Mid-State Oil in 1968, when Sunoco purchased Midstate. When Sunoco withdrew from the Southeast, Mid-State was spun off as an independent and continues to operate Travelers stations today.

MID-CONTINENT PETROLEUM - TULSA, OK: Mid-Continent was organized in 1916 from the various holdings of oil pioneer Josh Cosden and others. Gasoline marketing under the Diamond brandname began in about 1925 and by the late-1920s, Mid-Continent operated a large number of stations throughout the Central United States. A super-premium fuel "Diamond D-X" was introduced in 1933 and by 1940 stations were branded as "Diamond D-X" stations. After WW II the brandname became simply "D-X". In 1955 Mid-Continent merged with Sunray Oil Co. and the corporate name became Sunray DX Oil Company, revised in 1962 to DX-Sunray Oil Company. In 1968 DX-Sunray merged with Philadelphia based Sun Oil Company and many dual logo Sunoco/DX items were introduced at this time. DX globes had been phased out by the time of the merger. Shortly after the merger the Sunoco brandname replaced DX in states where Sunoco already marketed. The DX brand was retained in other areas until after 1980 when Sunoco began gradually replacing DX. As of this writing only small areas of DX branding remain, primarily jobber operated stations that just haven't yet been converted.

****SUNRAY OIL:** Sunray, a mid-continent area producer/refiner dating back to the 1920s, was historically very involved with production and unbranded marketing with very little attention paid to branded "Sunray" marketing. Gasoline was sold under the company name through a small number of jobbers and due to the limited branded marketing, the globes listed are rare. Sunray Oil merged with Mid-continent Petroleum in 1955 and the "DX" brand was added to Sunray logos at existing Sunray stations. In about 1962 those few stations simply became "DX" stations.

****PREMIER - LONGVIEW, TX:** Independent Texas refiner with stations in Texas, Oklahoma, Arkansas and Louisiana. In 1964 the company merged with Sunray DX. Some stations were spun off to Vickers, while DX retained the Premier name. Some unsuccessful DX stations were later converted to Premier and the brand survived at a small number of locations until DX rebranded Sunoco in the 1980s.

SUNOCO

****BARNSDALL REFINING CO. - BARNSDALE, OK:** One of the oldest oil refiners in the business, Barnsdall began retail gasoline marketing in 1921. At one time one of the largest mid-continent marketers, Barnsdall was reorganized during the Depression as Barnsdall Refining and in 1940 the Bareco brandname replaced the Barnsdall brand. In 1952 Bareco closed their last refinery to concentrate on wax production and the gasoline marketing was sold off to Mid-Continent Petroleum. Mid-Continent eventually phased out the Bareco brand after the merger with Sunray.

****MONARCH MANUFACTURING CO. COUNCIL BLUFFS, IA:** Owner of the famous "MonaMotor" trademark, Monarch began gasoline marketing in about 1920 and merged into Barnsdall in 1929.

WITCO: Lubricants refiner and marketer that produced motor oils under the Kendall and Amalie brandnames. In 1996 Sun Oil purchased the lubricants operations, and the Kendall and Amalie brands continue as industrial lubricants distributed by Sun Oil Company.

KENDALL OIL - BRADFORD, PA: Kendall Refining was founded as a lubricants manufacturer and refiner in Bradford, Pennsylvania in 1881. The company had been well established in motor oils when the first venture into gasoline came in 1922. The company supplied stations in Pennsylvania and New York until after the first gas shortage in 1973 when the decision was made to concentrate on lubricants. The last Kendall branded station closed in 1976. Kendall became a unit of Witco Chemical, refining and marketing motor oils and lubricants. In 1996 the Kendall (and Amalie) brands were sold to Sun Oil.

****AMALIE:** Old Pennsylvania lubricants brand purchased by Witco Chemical in the 1960s. Sold to Sun Oil in 1996.

TENNECO

****TENNECO - HOUSTON, TX:** Tenneco was formed in 1961 when Tennessee Gas Transmission (a pipeline company) merged with Bay Petroleum of Denver, Colorado. Bay had been in the process of buying out regional private brands throughout the Southeast.

****BAY PETROLEUM - DENVER, CO:** Bay was founded in McPherson, Kansas in 1935, although little is known about the early history. By the mid-1950s, the operation had shifted to Denver, Colorado and in 1961 Bay Petroleum merged with Tennessee Gas Transmission to form Tenneco. During the 1960s, Tenneco purchased several private brand operations that continued to operate "company direct". A short-lived network of branded Tenneco dealers was created but most of the "Tenneco" stations were branded "Bay". Bay, while still independent, had purchased and rebranded several southern marketers. When Tenneco decided to get out of retail gasoline marketing, the remaining portion of the Bay brand, mostly in Georgia and Florida, was sold to Dixie Oil of Tifton, Georgia, a Tenneco supplied independent that continues to operate Bay branded stations. A note of trivia here: The Bay shield logo, last revised about 1939, is the oldest logo ever to be associated with a major brand that is still in use today, without change.

****CITIZENS OIL COMPANY - TALLAHASSEE, FL:** Citizens had originally operated as a franchise of Atlanta based Greyhound stations and had introduced the Citizens 77 brand logo in 1937. In about 1950 Citizens merged with Atlanta based Speed Oil

TENNECO

Co. Citizens 77 operated dealer stations while the Speed brand was used in their direct operated stations. With the 1957 purchase by Bay Petroleum, the Speed name was retained for several years in a transitional Bay shaped "SPEED" shield.

****SPEED OIL CO. - ATLANTA, GA:** Atlanta based discounter that merged with Citizens Oil Company in about 1950.

****GULF COAST OIL CO. - NEW ORLEANS, LA:** Gulf Coast was bought out by Bay Petroleum in the late 1950s and rebranded Bay.

****TENNECO - HOUSTON, TX:** Tenneco brandname introduced in 1961 and used on a group of dealer operated "Class A" stations in the Southeast. In 1970 Tenneco began experimenting with combination gasoline/convenience stores and built a network of stores in southern metropolitan markets, eventually expanding the store concept to their Direct and Red Diamond locations. In the late 1980s Tenneco sold off the gasoline marketing to TOC Retail and Tenneco branded locations were rebranded "Majik Market".

****DIRECT OIL CO. - NASHVILLE, TN:** Direct Oil Co. was founded in Nashville in 1946 by Calvin Houghland, the son of John Mason Houghland, founder of Spur Distributing. In the early years Direct acted much like a Spur jobber, distributing products through stations in eight states. Tenneco purchased the operation, totalling 132 stations, in 1962 and as of this writing, TOC retail, successors to the Tenneco marketing, continue to operate Direct stations throughout the Southeast.

****RED DIAMOND OIL COMPANY - PICKENS, SC:** Trackside discounter that was operating about 25 stations in North and South Carolina when purchased by Tenneco in the 1970s. Most of the original Red Diamond locations remain branded Red Diamond today.

****DIXIE OIL CO. - TIFTON, GA:** Dixie Oil Co. of Tifton, Georgia is an independent company supplied by Tenneco beginning in the late 1960s. When Tenneco got out of retail gasoline marketing in the late 1980s, Dixie Oil purchased most of the former Bay branded operation in Georgia and Florida and continues to operate the Bay stations today.

****SUNSET:** Sunset is a Dixie secondary operation.

****Q-S PETROLEUM - NASHVILLE, TN:** Regional independent purchased by Tenneco in the 1970s. They operate under the "Trust" brandname.

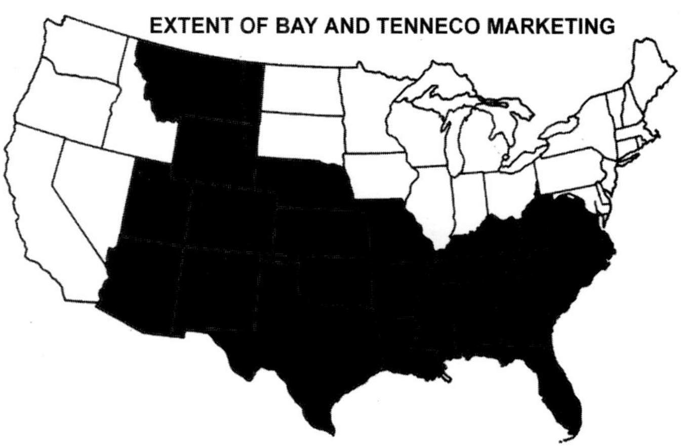

TENNECO
HOUSTON, TX

1962-1988

BAY PETROLEUM
DENVER, COLORADO

1930s-1940

1940-PRESENT

TENNECO
SECONDARY BRAND

1970s-1988

GULF COAST OIL CO.
NEW ORLEANS, LA

1930s-1955

DIRECT OIL CO., NASHVILLE, TENNESSEE

1946-1962

1962-1993

1993-PRESENT

DIXIE OIL CO.
TIFTON, GA

1947-1960

1950s-PRESENT

DIXIE SECONDARY BRANDS

1950s

TENNECO

GREYHOUND OIL COMPANY, ATLANTA, GA

PREWAR

POSTWAR

CITIZENS OIL CO., TALLAHASSEE, FL

1946-1954

1954-1957

1955-1957

SPEED OIL COMPANY, ATLANTA, GA

1938-1946

1946-1957

1957-1962

RED DIAMOND OIL CO.
PICKENS, SC

1940s-PRESENT

1970s-PRESENT

QS PETROLEUM
NASHVILLE, TN

1960s-PRESENT

E-Z SERVE, HOUSTON, TX

1970s

1980s-PRESENT

1970s-PRESENT

THE TEXAS COMPANY

1907-1920s | 1915-1926 | 1926-1936 | 1936-1944

HAVOLINE OIL COMPANY NEW YORK, NY

1906-1918

INDIAN REFINING COMPANY LAWRENCEVILLE, IL

1910-1923 | 1923-1934

****THE TEXAS COMPANY - NEW YORK, NY:** Throughout this country and around the world the first gasoline pump most people ever saw was a Texaco pump. Founded in 1902 in Port Arthur, Texas, Texaco was early on a worldwide producer, refiner and marketer. Their early "sales offices", set up to market oils and greases, specialty products, asphalts and even roofing shingles were to become the first organized network of gasoline filling stations. Promoting gasoline very heavily before 1910, they became the first marketer to establish a brand image -- in 1915 -- and went on to become, with the 1928 purchase of California Petroleum, the first oil company to operate gas stations in all 48 (and later all 50) states. They continued to build and operate stations everywhere and with the coming of the Interstate Highway System in 1956 and the many new locations offered by it, Texaco eventually branded more than 40,000 gas stations in the United States. The oil shortages of the 1970s forced Texaco to pull back, entirely moving out of some areas of the country by the mid-1980s. Texaco's purchase of Getty in 1984 added some former Skelly locations in areas Texaco had withdrawn from, but forced the company to the edge of bankruptcy when Pennzoil sued over the Getty purchase and won. Forced to scale back even more and to team up with a group of overseas investors, Texaco survived and currently markets through more than 15,000 stations in 48 states. Eastern Hemisphere marketing is operated under the "Caltex" brandname, created in the 1936 establishment of Anglo-American Oil Company, controlled by equal partners Texaco and Chevron.

****INDIAN REFINING COMPANY - LAWRENCEVILLE, IL:** Indian was a refiner/marketer with branded gasoline stations in 19 Northeast and North Central states. They are best remembered for the motor oil brand Havoline, introduced in 1906 by Havoline Oil Co. of New York which Indian Refining purchased in the late 1910s. Texaco continues to use the Havoline brand today. Texaco purchased Indian in 1931 in order to acquire the Havoline process patents and Indian stations were rebranded Texaco, while "Indian Gas" became a Texaco economy grade gasoline marketed until 1942.

****McCOLL BROTHERS, LTD. - TORONTO, ONTARIO:** McColl Brothers was established in 1873 as a manufacturer of industrial lubricants. They expanded into gasoline marketing in 1916, primarily as a distributor, and in 1925 began operating a refinery for the production of their new gasoline, branded "Red Indian".

****FRONTENAC OIL REFINERIES, LTD. - MONTREAL, QUEBEC:** Established in October 1925, Frontenac was a reorganization of the earlier entity, Nations Oil Refineries, LTD. They introduced the Frontenac brandname shortly after the company was organized, and it remained the primary brandname used by Frontenac until 1932 when the McColl "Red Indian" brand was adopted.

****NATIONS OIL REFINERIES, LTD. - MONTREAL, QUEBEC:** Nations Oil Refineries, Ltd., was a refining and marketing company purchased by Frontenac Oil Refineries, Ltd., in the 1926 establishment of the company. They had operated un-

TEXACO REFINING AND MARKETING / STAR ENTERPRISE

1936-1968

1965-1982

1982-PRESENT

CALIFRONIA PETROLEUM COMPANY

PRE-1928

ANGLO-AMERICAN OIL COMPANY

1936-1939

1939-PRESENT

McCOLL-FRONTENAC LTD., MONTREAL, PQ

1927-1939

1939-1947

REGENT REFINING, PORT CREDIT, OT

PRE-1957

der the Cyclo brandname for several years, but the Frontenac brandname soon replaced Cyclo.

McCOLL-FRONTENAC OIL, LTD., MONTREAL, QUEBEC: After the merger of McColl Bros. and Frontenac Oil Company in November 1927, the two entities continued to operate separately until 1932, when the operations were completely merged. The merged entity continued to market under the "Marathon" brandname, using the red Indian logo. The brandname later became representative of the logo "Red Indian" In the late 1930s Texaco began investing in McColl-Frontenac, and in return McColl Frontenac purchased, in 1939, Texaco's operations that had been established in Western Canada. The unique name and image continued to be used through the gradual phasing in of Texaco brandnames, which began in 1941. The final Red Indian signs came down in 1947. Increased investment by Texaco led to the corporate name change in 1959, when McColl-Frontenac became Texaco Canada, LTD. Operations continued under the Texaco corporate name and brandnames until 1990, when Imperial Oil, the Canadian Exxon affiliate, purchased Texaco Canada. The Esso brandname replaced Texaco at hundreds of stations nationwide, although Texaco lubricants continue to be available for the automotive aftermarket.

REGENT REFINING, LTD. - PORT CREDIT, ONTARIO: Regent was an Ontario based refiner-marketer that McColl-Frontenac purchased in 1957. Stations were rebranded Texaco.

TOTAL

LEONARD REFINERIES, ALMA, MI

1935-1938

1938-1941

1941-1950

1950-EXP

1950-1957

1957-1964

1964-1970

ROOSEVELT OIL COMPANY, MT. PLEASANT, MI

1920s-1930s

1930s-1946

1946-1955

MID-WEST REFINING COMPANY, MT. PLEASANT, MI

1934-1946

1946-1970

CAVALIER SERVICE STATIONS, FLINT, MI

1950s-1968

1968-1990s

ATOMIC POWER, MIDWEST SECONDARY BRAND

1950s

TOTAL

TOTAL PETROLEUM, ALMA, MI AND DENVER, CO

TOTAL — 1970-TEMP

TOTAL — 1970-1976

TOTAL — 1976-1992

TOTAL — 1989-PRESENT

COL-TEX REFINING COMPANY, OKLAHOMA CITY, OK

COL-TEX GASOLINE — 1931-1946

COL-TEX Challenge GASOLINE — 1946-1953

ANDERSON-PRICHARD OIL, OKLAHOMA CITY, OK

ANDERSON MOTOR FUELS PRICHARD — 1930s-1946

ANDERSON Challenge PRICHARD — 1946-1953

APCO PETROLEUM PRODUCTS — 1953-1970

APCO — 1970-1989

Apco Total Petroleum, Inc. — 1989-1993

LESH REFINING, ARKANSAS CITY, KS

Lesh's — 1910s-1928

KANOTEX REFINING, ARKANSAS CITY, KS

KANOTEX GASOLINE — 1909-1928

KANOTEX — 1928-1953

TOTAL

VICKERS PETROLEUM CORPORATION, WICHITA, KS

1930s

1938-1949

1949-1968

1968-1985

1985-PRESENT

1980s CORPORATE

TRUMAN ARNOLD COMPANY
TEXARKANA, TX

1960s-PRESENT

BEN FRANKLIN REFINING
ARDMORE, OK

1940s

KENT OIL COMPANY
SALINA, KS

1930s-1950s

BELL OIL AND REFINING
TULSA, OK

1930s-1964

EXTENT OF
TOTAL BRANDS

KNIGHT OIL COMPANY
SPRINGFIELD, MO

1950s-1960s

TOTAL

TOTAL PETROLEUM - ALMA, MI / DENVER, CO: Total, a unit of the French government, entered the US market in 1970 with the purchase of Leonard Refineries. The more than 700 Leonard stations in Michigan were rebranded Total by 1971 and the company remained a one-state marketer until 1978 when they purchased APCO. They have since acquired Vickers Petroleum, from Esmark in 1980, and the Road Runner stations of the Truman Arnold Corp. Currently they market in Michigan and surrounding states under the Total brand, and throughout the Midwest under Total, Vickers and Apco. In 1997, they announced plans to merge with Ultramar Diamond Shamrock, with UDS remaining the dominant company, retaining use of all three brandnames, Ultramar, Diamond Shamrock, and Total.

****McCLANAHAN OIL COMPANY - ALMA, MI:** Founded in the early 1930s in Michigan's oil region. McClanahan entered the gasoline market in 1935 and the company was reorganized as Leonard Refineries in 1936. Gasoline was rebranded Leonard in 1938.

****LEONARD REFINERIES - ALMA, MI:** McClanahan Oil Company was reorganized in 1936 as Leonard Refineries. The Leonard gasoline brand was introduced in 1938. In 1970 Total Petroleum purchased Leonard Refineries and rebranded the stations "TOTAL". They continue to operate under the Total brand in Michigan, Ohio and Wisconsin.

****ROOSEVELT OIL CO. - MT. PLEASANT, MI:** Michigan marketer operating under the "ROC" and Roosevelt brands before WWII and branding Power-Flight from 1946 until 1955. Purchased by and rebranded Leonard in 1955.

****MID-WEST REFINING - MT. PLEASANT, MI:** Michigan refiner/marketer that marketed under the Mid-West brand until after WWII when they licensed the "White Rose" trademark from National Refining. Stations were then rebranded Mid-West White Rose. Leonard purchased Mid-West in 1955 and continued to operate the White Rose stations as a secondary brand until after Total purchased Leonard in 1970. Stations were then rebranded Total.

****CAVALIER SERVICE STAS, INC. - FLINT, MI:** Michigan regional marketer, Cavalier sold out to Leonard Refineries in 1967. In 1968 the operation was rebranded "BEST" and when Leonard sold out to Total in 1970, Best was retained as a secondary brand. Gradually changed to Total with only three Best stations in operation as of this writing.

****ANDERSON PRICHARD - OKLAHOMA CITY, OK:** Founded in Oklahoma City in 1921, Anderson-Prichard was marketing gasoline under the Col-Tex brand in several Midwestern states prior to 1930. In 1946 the Col-Tex stations in Oklahoma were rebranded Anderson-Prichard while the remaining territory continued to use the Col-Tex brand. In 1952 Anderson-Prichard purchased the extensive Kanotex marketing and introduced the APCO brand to replace Kanotex and Anderson-Prichard. The Col-Tex operation was sold off to Cosden, who continued to operate Col-Tex as a secondary brand until their merger with Fina in 1963. Control of APCO passed to Ashland Oil in the late 1950s and in 1978 Total Petroleum purchased the Apco operation. Total continues to brand a few dealer operated APCO stations today, but most have rebranded Total.

****KANOTEX - ARKANSAS CITY, KS:** Founded as Superior Refining in Longton, KS in 1905, the Kanotex trademark was first used in 1909. The company moved to Arkansas City, KS in 1917 and later merged with Arkansas City based Lesh Refining. Innovative marketers, the KANOTEX trademark appeared in more that 20 Midwestern states. In 1952 Anderson Prichard purchased Kanotex and rebranded the Kanotex operation, as well as their own Anderson-Prichard stations to APCO.

****LESH REFINING CO. - ARKANSAS CITY, KS:** Early Kansas refiner that merged with Kanotex in 1928.

****VICKERS PETROLEUM CORPORATION - WICHITA, KS:** Founded in 1917, Vickers was a mid-continent refiner/marketer, operating stations in the central United States. The company remained in the Vickers family until 1968 when it was purchased by Esmark, the conglomerate that had its roots in Swift and Company, the Chicago meat packers. Under Esmark, the brand expanded into about a dozen Midwestern states. Total purchased Vickers in 1980 and continues to operate a few stations under the Vickers brand today, as most have rebranded to Total.

****KNIGHT OIL CO. - SPRINGFIELD, MO:** Missouri marketer purchased by Vickers in about 1959. Rebranded Vickers in the early 1960s.

****BELL OIL AND GAS - TULSA, OK:** Independent refiner/marketer, founded in 1918, that was purchased by Vickers in the late 1960s. Bell was primarily a wholesale marketer with only a few branded stations.

****BEN FRANKLIN REFINING CO. - ARDMORE, OK:** Oklahoma refiner that operated a few Ben Franklin branded stations in the 1930s. Merged with Bell Oil prior to WWII.

****KENT OIL COMPANY - SALINA, KS:** Large regional Vickers jobber that was purchased by Vickers and rebranded in the 1960s. Originally operated under the "Kent" and Jayhawk" brandnames.

United Refining

United Refining
Company

CORPORATE LOGO INTRODUCED 1974

UNITED REFINING, WARREN, PA

Vitalized KEYSTONE GASOLINE — 1930s-1954

KEYSTONE — 1954-1980s

KEYSTONE — 1980s

KEYSTONE — 1992-PRESENT

MINUTE MAN SERVICE, INC.
LAKEWOOD, NY

MINUTE MAN — 1960s-1972

minute man — 1972-1980s

BEAVER GASOLINE CO.
FREEDOM, PA

Falcon — 1960s-1980s

SKAT OIL CO.
TITUSVILLE, PA

SKAT OIL COMPANY — 1960s-1970s

UNITED OIL MARKETING
ERIE, PA

UNITED — 1926-1975

EXTENT OF
UNITED MARKETING

United Refining

UNITED REFINING CONVENIENCE STORES

1980s-PRESENT

SEAWAY STATIONS
ROCHESTER, NY

1960s-1973

ELK REFINING
CHARLESTON, WV

1920s-1930s

1940s-1963

OSCEOLA REFINING
REED CITY, MI

1946-1960

1960-1989

****UNITED REFINING COMPANY - WARREN, PA:** Founded in 1902 as a production company, United's strength in the early years was as a lubricants manufacturer. In the late 1920s they created a subsidiary, Emblem Oil Company, and entered gasoline marketing through Emblem under the Keystone brandname. An affiliated company, Elk Refining of Charleston, WV, also began as a lubricants manufacturer and was marketing gasoline in West Virginia under the Keystone brand as well. Control of Elk Refining was sold to Pennzoil in 1952, the balance in 1963. In 1953 United consolidated its holdings to focus on branded gasoline marketing and the Keystone brand. During the 1960s United concentrated on acquiring a series of private brands to strengthen their position in a price-conscious market. During the 1970s gas shortages much of United was disassembled, but the remaining company saw a revitalization in the 1980s in the convenience store business under the Red Apple and Kwik Fill brands. The 1990s saw another revitalization of the Keystone brand, which continues to be seen along Pennsylvania and New York highways today.

****ELK REFINING COMPANY - CHARLESTON, WV:** Refiner and marketer established by members of the same family that founded United Refining. Elk marketed primarily lubricants in the early years, although some gasoline was sold under first the "Elk" and later "Keystone" brandnames. Control of Elk passed to Pennzoil in 1952, the balance in 1963, and Keystone stations then became Pennzoil stations.

****MINUTE MAN SERVICE, INC. - LAKEWOOD, NY:** Regional discounter purchased by United Refining in the 1960s. Most stations were converted to United's Red Apple or Kwik Fill brands in the 1980s.

****SKAT OIL COMPANY - TITUSVILLE, PA:** Regional discounter purchased by United Refining in the 1960s. Most stations were converted to United's Red Apple or Kwik Fill brands in the 1980s.

****BEAVER GASOLINE COMPANY - FREEDOM, PA:** Regional discounter operating stations under the "Falcon" brand, purchased by United Refining in the 1960s. Most stations were converted to United's Red Apple or Kwik Fill brands in the 1980s.

****UNITED OIL MARKETING - ERIE, PA:** Regional terminal company and marketer that dated from the 1920s, United Oil Marketing was completely unaffiliated with United Refining until purchased by them in 1968. Operated a small number of stations under their own United brand into the 1970s.

****KWIK FILL and RED APPLE:** United's convenience store operations are handled under these two brandnames. Kwik Fill was originally an Ohio regional marketer.

****SEAWAY STATIONS, INC. - ROCHESTER, NY:** Small New York private brand purchased by United in 1969. Brandname was discontinued almost immediately after the purchase.

****OSCEOLA REFINING COMPANY - REED CITY, MI:** Small Michigan refiner marketer purchased by United in 1973. Following the gas shortages, and a fire at the refinery in the 1970s, Osceola was sold to Michigan's Naph-Sol refining, which kept the brandname alive into the 1980s.

TOSCO - THE OIL SHALE CO.
UNION OIL COMPANY OF CALIFORNIA

1913-1932

1932-1938

1938-1947

1947-1955

1955-1967

1967-PRESENT

1967-1983

1983-PRESENT

THE PURE OIL COMPANY USA

1954-1970 and 1992-PRESENT

HARBOR REFINING COMPANY - LONG BEACH, CA

(UNION SECONDARY BRAND)

1940s

****UNION OIL CO. OF CALIFORNIA:** Organized in 1890 in Santa Paula, California from the holdings of founders Lyman Stewart and Wallace Hardison and others, Union began selling California made gasoline to California motorists in 1913. Throughout its history a dominant West Coast marketer, Union was operating stations in 12 states at the time of the 1965 merger with Chicago based Pure Oil. By 1970 the Union 76 brand had replaced Pure and the company was marketing coast-to-coast. The corporate name became Unocal Corp in 1983 and the brand became Unocal 76 in 1985. In 1992 Unocal elected to withdraw from branded marketing in the Southeastern United States. Former Unocal jobbers banded together and resurrected the Pure brandname, and Pure stations, from a base in Savannah, Georgia are found in 10 states today. In 1997 Unocal sold off their marketing division, "76 Products Company" and the "76" brand to TOSCO. TOSCO continues to operate stations under the Unocal 76 brand.

****LION OIL CO. - EL DORADO, ARKANSAS:** Lion Oil Co. was founded in 1921 in the South Arkansas oil fields. Primarily an exploration/drilling company and an early innovator in petrochemicals, Lion was operating a few service stations by the early 1920s. From the 1920s until the early 1950s, gasoline marketing was limited to Arkansas, Mississippi, Western Tennessee and Northern Alabama and Louisiana, although Lion Naturalube Motor Oils were distributed throughout the U.S. In 1954 the Monsanto Chemical Corporation purchased Lion and began to

TOSCO - THE OIL SHALE CO
LION OIL CO., ELDORADO, ARKANSAS

1920s-1959

1920s-1959

1930s-1959

1959-1965

1965-1981

CIRCLE K CORPORATION, PHOENIX, AR

1958-1977

1977-PRESENT

expand the marketing territory throughout the SouthCentral U.S., expanding as far west as central Kansas and as far north and east as Bristol, Virginia. In 1959 Lion introduced the jumping lion logo, replacing the "lion and mound" logo that had been in use since the 1920s. During the 1960s the Monsanto name appeared on gas pump signage along with the Lion brand and some St. Louis, Missouri area stations added Monsanto signage alongside the Lion signs, creating speculation that the Lion name would be replaced. The experiment was short lived as Monsanto decided to get out of retail gasoline and in 1971 Monsanto sold the petroleum divisions of Lion to "The Oil Shale Co." (TOSCO), retaining the petrochemical divisions. Lion branded marketing was discontinued in 1981, although TOSCO/Lion is still involved in unbranded marketing.

****TOSCO - LOS ANGELES, CA:** An accumulation of refining and marketing capability, Tosco assumed control of Lion Oil in 1971. In 1981 they discontinued branded marketing, only to re-enter retail markets in the 1980s under a franchise of the BP brand. In 1996 they purchased Circle K stores of Phoenix, Arizona, a coast-to-coast operator of convenience stores. In 1997 they purchased 76 Products Company, and are in the process of absorbing the remaining operations of the Unocal 76 brand.
****CIRCLE K - PHOENIX, AZ:** Founded in the late 1950s as a convenience store operator, Circle K has had a varied history through the years. Cycles of expansion were met with cycles of hard times. In 1996 Circle K was purchased by TOSCO, and Circle K stores continue to operate in select markets coast-to-coast today.

INDUSTRY SYMBOLS

LUBRIZOL CORP. - CLEVELAND, OH
(SOLVENIZED)

1935-1946

LUBRIGAS, INC.
CHICAGO, IL

1930s

PENNSYLVANIA GRADE CRUDE OIL ASSOCIATION

1926-1960s

SOCIETY OF INDEPENDENT GASOLINE MARKETERS OF AMERICA

1960s-PRESENT

INDEPENDENT OIL MENS' ASSOCIATION
CHICAGO, IL

1922-1970s

1926-1932

1926-1932

ETHYL CORPORATION, RICHMOND, VIRGINIA

1923-1924

1926-1941

1941-1960s

CANADA
1926-1941

INDEPENDENTS OF THE NORTHEAST AND EASTERN CANADA

CONTINENTAL REFINING CO.
OIL CITY, PENNSYLVANIA

1920s-1951

QUINCY OIL , BOSTON, MA

1920s-1980s

IRVING OIL CO., ST. JOHN, NEW BRUNSWICK

1926-1930s

1930s-1960

1960-PRESENT

PURE QUILL
ROCHESTER, NY

1930s-1940s

RARITAN OIL CO.
EDISON, NJ

1960s

SEARS OIL COMPANY - ROME, NY

SEARS

1920s-PRESENT

INDEPENDENTS OF THE SOUTHEAST

COLONIAL OIL INDUSTRIES, SAVANNAH, GA

1932-1946

1946-1956

1956-PRESENT

1959-PRESENT

1959-PRESENT

1982-PRESENT

COLONIAL OIL CO., JACKSONVILLE, FL

1932-1977

SAN ANN SERVICE BOAZ, AL

1960s-PRESENT

EAST COAST OIL COMPANY RICHMOND, VIRGINIA

1947-1960

1960-1977

1977-PRESENT

SHELBY PETROLEUM, LA FOLLETTE, TN

1939-1960s

1946-1960s

1960-1970s

1970s-PRESENT

INDEPENDENTS OF THE SOUTHEAST

PILOT OIL COMPANY, KNOXVILLE, TENNESSEE

1958-1968 1968-1977 1977-1985 1985-PRESENT

TAYLOR OIL COMPANY, WINSTON SALEM, NC

1940-1958 1958-1964 1964-1970s

1959-1963 1963-1967 1967-1982 1982-PRESENT

A. T. WILLIAMS OIL CO.
WINSTON SALEM, NC

1963-1975 1975-PRESENT

GATE OIL CO., JACKSONVILLE, FL

1963-1991 1991-PRESENT

INDEPENDENTS OF THE SOUTHEAST

L.M. HUGHEY COMPANIES TAMPA, FLORIDA

1967-1972

1972-1980s

1980s-1988

1986-PRESENT

H. L. HUNT OIL CO., DALLAS, TX

PUBLIX OIL CO. MORRISTOWN, TN

1939-1955

1955-1971

1971-PRESENT

1960s-1987

SERVICE DISTRIBUTING CO., ALBRMARLE, NC

1946-1964

1956-1980

1956-1980

1980-PRESENT

INDEPENDENTS OF THE SOUTHEAST

**SAVINGS DISTRIBUTING, D&H CORPORATION
TUPELO, MS**

1960s

1970s-PRESENT

**SOMERSET OIL CO.
SOMERSET, KY**

1970s-PRESENT

**SMILE OIL CO.
HICKORY, NC**

1955-1989

THORNTON OIL CO., NEW ALBANY, IN

1973-1980s

1973-1980s

1980s-PRESENT

**SOUTHLAND OIL CO
YAZOO CITY, MS**

1950-1967

1967-PRESENT

**SOUTHERN STATES OIL CO.
JACKSONVILLE, FL**

1950s

**TANKAR STATIONS
NORFOLK, VA**

1933-1979

THONI OIL CO. NASHVILLE, TN and MIAMI, FL

1930s

1940s-1991

INDEPENDENTS OF THE MIDWEST

BAY-DOW CHEMICAL
BAY CITY, MI

1950s-1980s

CERTIFIED OIL CO.
COLUMBUS, OH

1940s-1970

1970-PRESENT

CASEY'S GENERAL STORES
DES MOINES, IA

1970s-PRESENT

CRYSTAL FLASH PETROLEUM
INDIANAPOLIS, IN
GRAND RAPIDS, MI

1930s

POSTWAR-PRESENT

DRAKE REFINERY STATIONS, INC.

1950s

GLOBE OIL CO.
MCPHERSON, KS

1930s-1946

GIBBLE OIL COMPANY, CUSHING, OK

1930s-1946

1946-1980s

1980s-PRESENT

INDEPENDENTS OF THE MIDWEST

HICKS GAS & OIL
AURORA, IL

1939-1950

1950-1980s

1980s-PRESENT

GAS CITY, GARY, INDIANA

1960s

1970s-PRESENT

GAS STOP, CHICAGO, IL

1970s-PRESENT

BUY-WRIGHT SERVICES, VERNON HILLS, IL

1950-1991

1991-PRESENT

KEAN OIL CO., CHICAGO, IL

1930s-1960s

1960s-PRESENT

MILLER & HOLMES ST. PAUL, MINNESOTA

1931-1970

1970-PRESENT

INDEPENDENTS OF THE MIDWEST

KOCOLENE OIL CO., SEYMOUR, IN

1943-1950

1950-1988

1988-PRESENT

1988-PRESENT

SWIFTY OIL COMPANY, SEYMOUR, IN

1960s-PRESENT

ILLINOIS OIL CO. ROCK ISLAND, IL

1926-1941

QUIK TRIP STORES, TULSA, OK

1958-PRESENT

MARINE OIL COMPANY, ST. LOUIS, MO

1930s-1960s

1960s-1970s

1946-1958

1958-1980s

INDEPENDENTS OF THE MIDWEST

SPEEDWING CORPORATION, ALTON, IL

--1950s--

PLATOLENE 500, CARMI, ILLINOIS

1930s-1973 | 1960s-1980 | 1980-1996 | 1996-PRESENT

SITE OIL COMPANY, ST. LOUIS, MO

1940s | 1950 | 1950s-1995 | 1995-PRESENT

STAR SERVICE & PETROLEUM, CENTRALIA, IL

1946-1960 | 1960-1984 | 1984-1991

INDEPENDENTS OF THE MIDWEST

ALLIED CHEMICAL
HOUSTON, TX

1950s-1980s

1970s

WACO OIL CO.
COVINGTON, KY

1960s-PRESENT

TOWN PUMP
BUTTE, MT

1960-PRESENT

TRACKAGE OIL
MILWAUKEE, WI

1930s

US OIL
COMBINED LOCKS, WI

1970s-PRESENT

UNIVERSAL OIL
WICHITA, KS

1960s-1970s

WARELUBECO, INC., JACKSONVILLE, IL

1946-1970s

1970s

1980s

1990s

SECURITY OIL, WICHITA, KS

1930s-1980s

WESTLAND OIL CO., MINOT, ND

1930s-1960s

1960s

INDEPENDENTS OF THE SOUTHWEST AND WEST COAST

BLAKELY OIL CO.
PHOENIX, AZ

1960s

WHITING BROTHERS
PHOENIX, AZ

1950s-PRESENT

FLETCHER OIL CO.
BOISE, ID

1930s-PRESENT

RITTERS GAS STATIONS
AUSTIN, TX

1960s

TESORE PETROLEUM ANCHORAGE, ALASKA
INCLUDING THE FORMER SIOUX REFINING COMPANY, NEWCASTLE, WY

1960s-PRESENT

1930s-1970s

SUNLAND REFINING, FRESNO, CA

1920s-1950s

1950s-1980s

TIME OIL
LOS ANGELES, CA

1946-1960s

1960s-PRESENT

U-SAVE AUTOMATIC PETROLEUM, SANTA MONICA, CA

1970s

1980s

1990s

INDUSTRY SYMBOLS

****SOLVENIZED - THE LUBRIZOL COMPANY - CLEVELAND, OH:** Lubrizol is a leading manufacturer of petroleum additives. In the mid-1930s they introduced the procedure for adding an engine-cleaning solvent to gasoline appropriately called "Solvenized". Branded marketers adding the chemical included Pocahontas Oil, Hickcok Oil, The Pure Oil Company, United Oil Marketing, and Jenney Manufacturing. The trademarked character "Dirty Dan - The Carbon Man" was adopted as a marketing symbol and used in various forms by the affiliated Solvenized marketers until WWII.

****LUBRI-GAS - CHICAGO, IL:** Another chemical additive company, specializing in a lubricating additive for gasoline. Used the camel logo to identify gasolines treated with Lubri-Gas.

****PENNSYLVANIA GRADE CRUDE OIL ASSOCIATION - OIL CITY, PA:** Formed in 1926 as a trade association organized to protect the integrity of motor oils bearing the "Pennsylvania" name. Each marketer wishing to market an oil advertised as coming from Pennsylvania crude was required to obtain membership and a license number for using the trademark Pennsylvania.

****INDEPENDENT OIL MEN OF AMERICA - CHICAGO, IL:** Also formed in 1922 as an association of independent, unbranded oil marketers all across the country. Originally conceived as a trade association, IOMA adopted the eagle logo and the brandname "Red Hat" for use by member marketers. Although Standard (IN) was successful in restricting the use of the Red Hat trademark in 1932, IOMA survived for many years thereafter, with a limited number of member marketers utilizing the eagle trademark into the 1970s.

****SOCIETY OF INDEPENDENT GASOLINE MARKETERS OF AMERICA:** A trade association of indepenent gasoline marketers, commonly referred to as SIGMA. Actively engaged in representing members today.

****ETHYL CORPORATION - RICHMOND, VA:** Advances in automotive engine design saw major steps forward during the decade of 1910-1920. Early two-cylinder and four cylinder designs gradually gave way to six cylinder in-line engines. As early as 1915 Cadillac offered a V-8 engine in certain models. Engine development, however, was hampered by the lack of availability of higher octane motor fuels necessary in developing higher compression ratios. The petroleum industry experimented with a number of blends and additives to gasoline in an attempt to create a fuel of higher anti-knock capability. Of the early additives, a blend of benzene and gasoline, marketed as "BENZOL" offered the most promise. The first effort by a major gasoline marketer to introduce a higher octane gasoline came in 1922 when Standard Of Indiana introduced "Solite", a gasoline with higher specific gravity than Standard's "Red Crown". (The method for determining octane rating was not established until 1929.) In early 1923, the General Motors Research Labs at Dayton, OH began testing the addition of tetraethyl lead to gasoline. Marketing experimentation began when GM teamed with Dayton independent Refiners Oil Company to introduce an "ethylized" gas, blending tetraethyl lead with gasoline at the pump. The experiment was quite successful and in September 1923 GM teamed up with Standard of Indiana to introduce Ethyl gasoline on a large scale. The following August GM and Standard Oil of New Jersey formed Ethyl Gasoline Corporation to franchise the Ethyl additive to gasoline marketers worldwide. The ink had hardly dried on this agreement when lead poisoning deaths in the Ethyl plants brought about a moratorium of the sale of "ethylized" gasoline until further study of potential harmful effects could be determined. After over a year of study by the US Surgeon General, Ethyl gasoline was reintroduced, with some restrictions on handling of the tetraethyl lead additive. With the re-introduction of Ethyl, every refiner meeting the Ethyl franchise requirements quickly signed on. Those who were deemed ineligible, for one reason or another, quickly introduced their own premium grade products in order to compete with Ethyl. Thus, in a period of about two years, virtually every marketer began offering at least two grades of gasoline. Ethyl Gasoline Corporation became simply Ethyl Corporation in 1942, and Ethyl contributed greatly to the war effort during WWII. Ethyl continued to be the anti-knock additive of choice for many years, but by the early 1960s marketers had ceased to identify their fuels as Ethyl. The phase out of leaded fuels, beginning in 1971, spelled the end of the high-octane Ethyl gasoline.

INDEPENDENTS OF THE NORTHEAST

****CONTINENTAL REFINING COMPANY - OIL CITY, PA:** An independent refiner founded in 1885 that began marketing gasoline in the 1910s. Operated stations in the Northeast and in Eastern Canada until closing in 1951.

****QUINCY OIL COMPANY - QUINCY, MA:** An 18th century trading company, Quincy expanded into gasoline marketing about 1920. A charter member of IOMA, they marketed under the IOMA eagle well into the 1970s. The company abandonded gasoline marketing in the early 1980s, but survives in other ventures today.

****IRVING OIL COMPAY - ST. JOHN, NEW BRUNSWICK:** Founded in the middle 1920s, Irving is Canada's only significant surviving independent oil company. In recent years they have expanded their marketing from the base in Eastern Canada into the New England states.

****PURE QUILL STATIONS - ROCHESTER, NY:** An early Rochester based operator of discount service stations.

****RARITAN OIL COMPANY - EDISON, NJ:** A mid-1960s operator of discount service stations.

****SEARS OIL COMPANY - ROME, NY:** Sears is an old local marketer, and continues to brand stations under the Sears name in New York. Despite occasional confusion, the company is in no way connected with Sears and Roebuck.

INDEPENDENTS OF THE SOUTHEAST

****COLONIAL OIL INDUSTRIES - SAVANNAH, GA:** Founded in 1921 as American Oil Company, Colonial was originally a gasoline marketer, branding to Pure in the 1930s. The name changed to Colonial Oil Company in 1932, and to Colonial Oil Industries in 1950. Branded gasoline marketing began in earnest in the early 1950s, and a chain of company owned and operated stations, branded Interstate, began in 1959. In 1982 Colonial introduced the brandname "Enmark", and began revamping the former Interstate stations into Enmark pumpers and convenience stores. They operate stores in the southeast today.

****COLONIAL OIL COMPANY - JACKSONVILLE, FL:** Founded in 1932 as an independent marketer, Colonial sold gasoline under the "Colonial Minuteman" brandname in Florida and Southern Georgia. Never a large operation, Colonial sold their stations to U-Save Automatic (USA Petroleum) in 1977.

****EAST COAST OIL COMPANY - RICHMOND, VA:** In the spring of 1941, Herbert Lomas, site surveyer for Southern Oil Stores (Dixie Vim) visited Richmond Virginia in search of locations to build Dixie Vim stations. While meeting with Franklin Robins, a local Pure dealer with several real estate interests, both men shared the interest in establishing a chain of discount service stations patterned after the successful Dixie Vim operation. A site was chosen on Richmond's discounter row, Hull Street, and a station branded "Big Tank" was opened in July 1941. In 1947 they purchased another local operation that branded their stations "East Coast", and the consolidated company adopted the East Coast brandname. They operate around 50 stations throughout Virginia today.

****SHELBY PETROLEUM - LaFOLLETTE, TN:** Founded in about 1938 as Fleet Oil Company, a chain of cottage style discount stations was established in Tennessee, Kentucky and Virginia. In 1959 Fleet Oil merged with another local marketer operating under the Sail Oil Company brandname. Fleet stations were rebranded Peer, as they continue to do today. Less than 20 stations remain in Tennessee and Kentucky.

****PILOT OIL CORPORATION - KNOXVILLE, TN:** Founded in 1958 with a single station in Weber City, VA, Pilot built a chain of discount stations in Virginia, Tennessee, and North Carolina during the the 1960s and 1970s. The success attracted the attention of Emro Marketing, which purchased part of Pilot in the 1980s. In about 1990 Pilot traded a group of service stations to Emro to regain its freedom, and began concentrating on building truck stops. Today they continue to operate a chain of discount stations in Virginia and Tennessee, along with truck stops and travel centers coast-to-coast.

****TAYLOR OIL COMPANY - WINSTON-SALEM, NC:** Taylor Oil, founded in 1930, was the branded Richfield jobber in Winston-Salem, NC when at the end of WWII they established a chain of discount gas stations they branded Travelers. The Travelers chain grew steadily in the 1950s, and in 1960 a companion brand, Etna, was introduced. During the early 1960s the Travelers stations were sold to Mid-State Oil Company of Salisbury, and Mid-State was purchased by Sun Oil in 1968, shifting operations to Lexington, NC. When Sunoco withdrew from branded marketing in 1992, Mid-State was sold off as an independent jobber and continues to market gasoline under the Travelers brand across North Carolina. The Etna operation proved very successful, growing steadily in the 1960s and 1970s, and today Taylor Oil, under the Etna brand, operates stations in North Carolina and Virginia.

****A.T. WILLIAMS OIL COMPANY - WINSTON-SALEM, NC:** During the late 1950s, A. Tab Williams began his career with Taylor Oil Company in Winston-Salem. Eventually he married Roby Taylor's daughter, and in 1963 launched the affiliated A. T. Williams Oil Company, operating stations under the Wilco brandname. After the gas shortages of the 1970s, the Wilco chain saw progressive growth, and today operates Wilco branded stations and travel centers, as well as serving as a branded jobber for Texaco, Exxon, and Shell, dual branding in many locations.

****GATE PETROLEUM - JACKSONVILLE, FL:** Founded in about 1963 by Herb Peyton, formerly a Billups Eastern executive who left Billups Eastern when the brand was acquired by Hess. Through the late 1960s Gate stations were added all through the South, and today the company has stations all through the Southeast.

****L. M. HUGHEY COMPANIES - TAMPA, FL:** Hughey is probably the most unknown operation among independent gasoline marketers. We believe the company was founded in the middle 1960s, and by 1968 Imperial stations were found throughout Virginia. Other brandnames, Eastern, Cargo, Charge, and Kash-N-Karry, appeared on stations from Florida north through Kentucky. In the late 1970s Hughey stations gradually converted to "Joy Food Stores", and in 1987 Derby signage appeared at many Hughey stations. Shortly thereafter, virtually all of the Hughey stations converted to the Coastal brand, remaining Coastal until the mid-1990s, when gradually the stations were rebranded back to Joy Foods and Imperial. Newly constructed sites of the 1990s also use the SuperTest brand name.

****H.L. HUNT OIL COMPANY - DALLAS, TX:** A unit of Hunt Refining, Parade stations first appeared in 1939. Hunt operated the Parade stations all across the South until the 1960s, when an Alabama jobber, Romaco, with ties to Hunt Oil bought out the locations and the rights to the Parade name, which is still in use at a limited number of locations.

****PUBLIX OIL COMPANY - MORRISTOWN, TN**: Publix began as Morristown's Sinclair jobber but lost its branding in the Arco-Sinclair merger of 1970. Overnight former Sinclair stations became Publix stations, the name was franchised across several Southeast states to other independent jobbers. Publix Oil of Morristown went out of business in about 1989, although occasional signs are still seen along East Tennessee highways.

****SERVICE DISTRIBUTING COMPANY - ALBEMARLE, NC:** Founded in 1933 by James Helderman, Service Distributing was originally a jobber for Atlantic. In the mid-1930s Service began building a chain of discount service stations across North Carolina. The operation saw explosive growth following WWII, and survived the gas shortage intact, helped by a guaranteed source of supply from Hess after the Helderman and Hess families became friends. In the early 1980s they began using the "Servco" brand, and began converting existing locations to convenience stores. They operate nearly 100 stations across North Carolina today, continuing close ties to Amerada Hess.

****SAVINGS DISTRIBUTING / D and H CORPORATION - TUPELO, MS:**
Postwar discounter that operates a far-flung chain of low-budget service stations, along with a growing chain of convenience stores operating under the "Dodge's Store" brand. They have about 50 stations today.

****SMILE OIL COMPANY - HICKORY, NC:** The Barringer family, Cities Service jobbers in Hickory, NC and surrounding towns, formed Blue Ridge Oil Company, later Smile Oil Company, to operate a chain of discount service stations appropriately branded Smile. Nearly 30 stations were in operation across Western North Carolina when the 1970s gas shortages began taking their toll on the Smile operation. Following a bad investment in developing a motor oil vending machine for use at unattended service stations then popular, Smile was forced to the edge of ruin. Pressures were building, and then Mr. Barringer was found dead in the Catawba River, an apparent suicide. The last branded Smile stations closed in 1981.

****SOMERSET OIL COMPANY - SOMERSET, KY:** Independent refiner and marketer, founded in 1932, that supplies a small number of stations in Eastern Kentucky and Tennessee.

****THORNTON OIL CO. - NEW ALBANY, IN:** The latest creation of James Thornton, who began his independent gasoline marketing career with Dance Oil Service in the Louisville area.

In the late 1940s Thornton founded Payless, and operated that company until the early 1970s when he sold most of the operation to Ashland. With a few stations remaining, Thornton began again, creating the Thornton Oil Company chain of service stations like Payless billed as being "From Canada to the Gulf Coast". In the early 1980s Thornton pulled back to core markets and continues to operate dozens of convenience stores today.

****SOUTHLAND OIL COMPANY - YAZOO CITY, MS:** Founded in 1939 as a production company and asphalt manufacturer, Southland entered gasoline marketing in 1950. The operation grew in the boom-times of the 1960s, but the gas shortages took their toll and today Southland operates only a very small number of stations in central Mississippi.

****SOUTHERN STATES OIL COMPANY - JACKSONVILLE, FL:** An independent Florida marketer from the 1940s through the 1960s.

****TANKAR STATIONS - NORFOLK, VA:** Founded by Bernard Solomon in Sheffield, AL in 1933, Tankar shifted the base of their operations to Norfolk, VA in 1939 following the success of stations opened in that area. Beginning in 1940, Tankar began constructing service stations in a unique design from converted rail tank cars. About 15 of the fantastic stations were in operation in Alabama, Virginia, and West Virginia when, in 1966, Bernard Stanton's son (he had changed his name to Stanton in the 1940s) joined him in the business. The younger Mr. Stanton began operating Robo automatic car washes at various Tankar station sites. By the early 1970s Tankar had closed many stations, converting the properties to car wash facilities. The last of the fantastic tank car constructed stations, in Portsmouth, VA, closed in 1979. A single pumper station, unbranded, remains in operation in conjunction with one Norfolk car wash site today, and the company continues to operate a successful chain of Robo automatic car washes.

****THONI OIL MAGIC BENZOL GAS STATIONS, INC. - NASHVILLE, TN and MIAMI, FL:** Founded in Nashville in the early 1930s as Magic Benzol Gas Stations, founder Richard Thoni purchased rail car lots of benzol blend gasoline and reselling it through economically constructed stations on highways leading into Southern cities. In the 1950s the company phased out the "Magic Benzol" name and image, and stations were simply branded Thoni. With the death of Richard Thoni in about 1990, the stations, numbering about 40 at the time, closed abruptly and never reopened.

INDEPENDENTS OF THE MIDWEST

****CASEY'S GENERAL STORES - DES MOINES, IA:** Founded in 1968, Casey's was an early convenience store operation that entered gasoline marketing in the early 1970s. They have steadily expanded, and now operate hundreds of stores throughout the Midwest.

****BAY OIL - DOW CHEMICAL CORP. - BAY CITY, MI:** Bay is believed to date from the 1950s, as a petroleum retailing division of Dow Chemical. The gasoline operation was discontinued in the 1980s.

****CERTIFIED OIL COMPANY - COLUMBUS, OH:** Dating from the 1930s and a statewide operator of discount service stations, Certified continues to brand over 100 locations in Ohio and West Virginia today.

****CRYSTAL FLASH PETROLEUM - INDIANAPOLIS, IN and GRAND RAPIDS, MI:** Crystal Flash is a family operation, founded in 1930 and operating from two bases, one in Indianapolis and one in Grand Rapids. In the postwar era they became a major independent marketer in their home territory, but have scaled back greatly in recent years and now operate less than 10 sites.

****DRAKES REFINERY STATIONS, INC. - LANSING, MI:** An innovative Michigan marketer, Drakes survives with less than 10 stations today.

****GIBBLE OIL COMPANY - CUSHING, OK:** Originally Cushing Oil, a refiner marketer based in Cushing, the company was reorganized in 1936 as Gibble Oil Company. Since that time the focus has been on simply retailing gasoline, which they continue to do in Oklahoma.

****GLOBE OIL AND REFINING - McPHERSON, KS:** Globe was a brand for the unbranded marketer. Almost operating on a franchise basis, it is similar in concept to the Dixie Distributors operation. Survived as a marketing company at least into the 1950s.

****HICKS GAS and OIL - AURORA, IL and ROBERTS, IL:** Consolidated in 1948 from a group of independent gas stations dating back to 1939, Hicks introduced the flamboyant Hicksatomic brandname and marketing style across Illinois in the 1950s. By the 1970s stations were simply branded Hicks, and several remain in operation today.

****GAS CITY - GARY, IN:** One of the later generation of independent gasoline retailers, Gas City operates stations and convenience stores in and around the Chicago area.

****GAS STOP - CHICAGO, IL:** Another Chicago area discounter, Gas Stop operates stations and convenience stores in and around the Chicago area.

****BUY-WRIGHT SERVICES, INC - VERNON HILLS, IL:** Founded in 1950, Buy-Wright Services sells gasoline at around 50 stations displaying the J&L brandname in the midwest and in Georgia.

****KEAN OIL COMPANY - CHICAGO, IL:** Founded in 1929, Kean is a Chicago area marketer.

****KOCOLENE OIL COMPANY - SEYMOUR, IN:** In 1938, Mrs. Carrie Myers purchased two service stations and became a branded jobber for Hoosier Petroleum, operating under the Hoosier Pete brandname. In 1943 the operation was rebranded Kocolene, as it continues to this day. In the 1960s family members established the neighboring Swifty Oil Company, also marketing gasoline in Indiana, Ohio, and Kentucky.

****SWIFTY OIL COMPANY - SEYMOUR, IN:** Founded in the 1960s by members of the Meyers family, operators of the Kocolene operation, Swifty brands in about 50 locations today.

****ILLINOIS OIL COMPANY - ROCK ISLAND, IL:** Founded in 1920, Illinois Oil began marketing gasoline in the 1920s and continued until the 1940s, when service stations were sold off to concentrate on the manufacturing of quality lubricants.

****MILLER AND HOLMES - ST. PAUL, MN:** Founded as a wholesale grocery company in 1896, Miller and Holmes entered gasoline marketing in 1931 with the establishment of a chain of service stations in cities across the upper Midwest. They continue to operate several dozen today.

****MARINE OIL COMPANY - ST. LOUIS, MO:** Marine Oil Company was a river terminal operator, buying gasoline at wholesale from a bustling terminal operation and reselling it to independent dealers under the Marine brand. Following WWII, Marine estabilshed a chain of company owned and operated stations coast to coast branded Mars. Mars stations were found in many large cities until the 1970s gasoline shortages eliminated a profitable source of supply. It is believed that much of the operation was sold to Murphy Oil.

****SPEEDWING CORPORATION - ALTON, IL:** An innovative postwar Midwestern independent operating under the Speedwing, Royal 99, and Pana brandnames.

****QUIK-TRIP STORES, INC. - TULSA, OK:** A pioneer convenience store operator, Quik Trip was founded in 1958 in Tulsa. Gasoline was added to many stores in the 1960s, and the company continues to operate stations in the Midwest and South today.

****PLATOLENE 500 - TERRE HAUTE, IN and CARMI, IL:** Originally a brandname of R. J. Oil Refining of Terre Haute, IN, the Platolene 500 brandname was introduced in the 1940s. In the early 1970s Crystal Oil purchased the refining operations, with the retail brand shifting operations to Carmi, IL. From that base they continue to operate stations today.

****SITE OIL COMPANY - CLAYTON, MO:** A family owned discount service station operator, Site stations are scattered throughout Midwestern cities.

****STAR SERVICE AND PETROLEUM - CENTRAILIA, IL and ST. LOUIS, MO:** An innovative but low budget discounter, Star Service dates from the 1930s and they operated a chain of nondescript discount service stations in over 20 states until 1993. Much like the Thoni operation, Star Service stations closed abruptly and never reopened.

****TEXGAS - ALLIED CHEMICAL - HOUSTON, TX:** Originally a regional marketer, Texgas became a part of Allied Chemical in the 1960s. They operated stations into the 1980s.

****WACO OIL COMPANY - COVINGTON, KY:** Founded in the Cincinnati suburbs in 1957, Waco is a discount marketer whose name has in recent years been obliterated as more and more sites branded. The company exists today as a branded marketer for Shell, Ashland, and Sunoco.

****TOWN PUMP - BUTTE, MT:** Regional marketer that has branded to Exxon in recent years.

****TRACKAGE OIL COMPANY - MILWAUKEE, WI:** One of many trackside discounters of the 1930s, Trackage seems to have been bought out or closed before WWII.

****U.S. OIL COMPANY - COMBINED LOCKS, WI:** A regional gasoline marketer founded in 1957, U.S. Oil operates dozens of stations today.

****UNIVERSAL OIL COMPANY - WICHITA, KS:** Dating back at least into the 1930s, Universal was a branded marketer in Kansas through at least the 1970s.

****WARELUBECO, INC. - JACKSONVILLE, IL:** Wareco has long been an innovative marketer with an enviable image in the marketplace. Founded in 1930 in Springfield, IL, Ware Oil, as it was known then, shifted operations to Jacksonville, IL in the late 1940s and introduced the Wareco brandname to a chain of discount service stations in Illinois, Indiana, and Missouri. Concentrating on customer service, Wareco adopted the image of the uniformed attendant as a logo. Following the evolution to self service later than most in the 1980s, Wareco's image was modernized as convenience stores were opened. In recent years the company shifted their image back to one of customer service, adopting again their attendant logo which is proudly displayed at stations today. Stations operate under the Wareco brand, as well as being branded Phillips 66 or Coastal.

****SECURITY OIL COMPANY - WICHITA, KS:** Little is known of this operation, only that the company marketed in Kansas under the Wings brandname from 1934 until the early 1980s.

****WESTLAND OIL COMPANY - MINOT, ND:** Westland is a regional refiner-marketer, dating from the 1920s, that operates stations under the Westland "Buffalo" brand in the northern Great Plains area.

INDEPENDENTS OF THE WEST and SOUTHWEST

****BLAKELY OIL COMPANY - PHOENIX, AZ:** Blakely was a postwar creation and a flamboyant marketer, operating stations in the growing southwest in the 1950s.

****FLETCHER OIL COMPANY - BOISE, ID:** An old time refiner-markter, Fletcher Oil continues to operate several stations under the "Veltex" brand.

****WHITING BROTHERS - PHOENIX, AZ:** Another gasoline marketing operation serving the desert southwest, Whiting Brothers stations were found at strategic intervals along through highways in Arizona and New Mexico. Several stations remain in operation to this day.

****RITTERS GAS STATIONS - AUSTIN, TX:** Texas regional marketer of the 1950s and 1960s.

****TESORO PETROLEUM - SAN ANTONIO, TX:** Founded as an exploration and production company in 1939, Tesoro entered retail marketing in 1968 with the purchase of Newcastle, Wyoming based Sioux Oil. Following the gas shortages, marketing was concentrated in Alaska, where some 40 stations are operated today.

****SUNLAND REFINING - FRESNO, CA:** An independent California refiner and marketer founded in 1929, Sunland continues to operate a small number of stations today, one of few surviving California independents.

****TIME OIL COMPANY - LOS ANGELES, CA:** A West Coast marketer that branded over 1,000 stations in the 1960s.

****U-SAVE AUTOMATIC, INC. (later USA PETROLEUM) - SANTA MONICA, CA:** Founded in the self serve boom of the late 1960s, U-Save Automatic became USA Petroleum in the 1970s. They continue to operate a small number of stations in California and in the Southeast today.

The authors of this book publish a monthly magazine dedicated to the preservation of petroleum collectibles and petroleum industry history. If you have additional information on any company marketing gasoline in the United States and Canada that should be included in future editions of this book, please contact us in care of:

PETROLEUM COLLECTIBLES MONTHLY
411 FOREST ST., LaGRANGE, OH 44050
(216) 355-6608 FAX (216) 355-4955

Interested in subscribing to the only complete monthly magazine serving the petroleum collectibles hobby? Contact us today! Visa, Mastercard, and Amex accepted.

NOTE: NEW AREA CODE (440) AFTER SEPTEMBER 1997

INDEX OF BRANDNAMES